Annual Conference of the American Forage and Grassland Council (AFGC 2020)

The More You Know, The Smarter You Grow

Greenville, South Carolina, USA
5 – 8 January 2020

ISBN: 978-1-7138-1448-1

Printed from e-media with permission by:

Curran Associates, Inc.
57 Morehouse Lane
Red Hook, NY 12571

American Forage and Grassland Council (AFGC)
PO Box 867
Berea, KY 40403
USA

Phone: 800.944.2342

www.afgc.org

Additional copies of this publication are available from:

Curran Associates, Inc.
57 Morehouse Lane
Red Hook, NY 12571 USA
Phone: 845-758-0400
Fax: 845-758-2633
Email: curran@proceedings.com
Web: www.proceedings.com

TABLE OF CONTENTS

GRAZING COVER CROPS: EFFECTS ON ANIMAL AND COVER CROP PERFORMANCE

R. C. Carrell[1], S. L. Dillard[1, 2], M. K. Mullenix[1], A. V. Gamble[2], R. B. Muntifering[1]

Cool-season cover crops can provide soil health benefits, as well as increased farm revenue when grazed with cattle. However, evidence is needed to verify the sustainability of grazing cover crops. Our objective was to determine animal and forage performance when grazing cool-season annual cover-crops in a peanut-cotton rotation. Twelve 3-acre pastures were established in a mixture of black oats, cereal rye, crimson clover, and T-raptor brassica to be grazed 0, 30, 60, or 90 days. Three tester steers were randomly placed in each paddock with the exception of control paddocks and allowed free choice grazing. Animals were weighed every 30 d to determine average daily gain (ADG) and total gain (TG). Forage was harvested bi-weekly and analyzed for yield, NDF and ADF using an ANKOM fiber analyzer (ANKOM Tech, Macedon, NY). All data was analyzed using the MIXED procedure of SAS version 9.4 (SAS Inst., Cary, NC). Forage yield showed differences between 0 and 60 days grazed (3,459 lbs/acre vs. 1,857 lbs/acre; $P = 0.01$), as well as between 0 and 90 days grazed (3,459 lbs/acre vs. 2,052 lbs/acre; $P = 0.04$). Differences were found in ADG between 90 and 60 days grazed (4.2 vs. 2.8 lbs/d; $P < 0.01$) and 90 and 30 days grazed (4.2 vs 2.7 lbs/d; $P < 0.01$). Differences in TG were detected between 90 and 60 days grazed (819 vs. 386.67 lbs; $P < 0.01$), between 90 and 30 days grazed (819 vs. 261.33 lbs; $P < 0.01$), and between 60 and 30 days grazed (386.67 vs. 261.33 lbs, $P < 0.01$). No differences in NDF (44.86%, $P = 0.99$) or ADF (27.20%, $P = 0.92$) were detected between treatments. These results indicate that different grazing periods could influence cattle growth and performance without greatly impacting cover crop yield or quality.

[1]Department of Animal Science, Auburn University, Auburn, rcc0057@auburn.edu; [2]Department of Crop, Soil, and Environmental Sciences, Auburn University, Auburn.

GRAZING MANAGEMENT CHOICES ON THE BEAVER CREEK FARM

Mike Jones[1]

We would like to show folks the choices we have made that help our cattle thrive on our property without feeding grain. Chemical fertilizer is used very little; grazing practices and unrolling of purchased hay add some nutrients to the soil. Sunlight, rain and grazing help build the soil in addition to the unrolled hay. In fescue trials by Alan Franzluebbers, we learned that because of good soil microbial activity, there is not enough increase in forage production to warrant using nitrogen fertilizer in August to stockpile grass for winter grazing. A brief history of tall fescue and how it became such an important grass in the southeast U.S.A. will be described. A fungus within the plant that produces a toxin that is detrimental to livestock is a problem to overcome. Grazing of different forages and rotational stocking help overcome the problem. Nitrogen fertilizer makes the toxins worse. We use cover crops, cool and warm-season native forages, legumes and even native grasses in fescue pastures. We use fire to increase quality of the native forages, and wildlife habitat is enhanced greatly. The years of working with NC State University, University of Georgia, NC Wildlife resources Commission, NC Forest Service, USDA-Natural Resources Conservation Service, American Forage and Grassland Council, National Grazing Lands Coalition and others have taught us many different techniques in grazing that help us make very good beef without grain. Water quality is much better because the soil is always covered.

[1]Farmer/Ecologist, Beaver Creek Farm, Surry County NC (bcangus@surry.net).
2010 Environmental Stewardship Award, Soil and Water Conservation Society, Hugh Hammond Bennett Chapter; 2013 Environmental Stewardship Award, North Carolina Forage and Grassland Council; 2019 Vice President, North Carolina Forage and Grassland Council.

GRAZING NATIVE WARM-SEASON FORBS AND GRASSES IN THE MID-SOUTH

J.D. Richwine and P.D. Keyser[1]

USDA cost-share programs place emphasis on high diversity blends to improve productivity and benefit pollinators and wildlife. Therefore, we conducted two native warm-season grass (NWSG) grazing studies at the AgResearch and Education Center in Greeneville, TN, from 2017-2019 to evaluate efficacy of establishment and persistence of an 11-species native forb/legume blend (FORB) when interseeded into existing big bluestem/indiangrass (BBIG) and switchgrass (SG) pastures. Forb/legume blend consisted of black-eyed susan, dixie ticktrefoil, false/oxeye sunflower, Illinois bundleflower, lanceleaf coreopsis, Maximilian sunflower, partridge pea, plains coreopsis, purple coneflower, purple prairie clover, and upright prairie coneflower. Each study was a completely randomized design. Five grazing treatments (no rest, early rest [no grazing May 29-Jun 19], middle rest [no grazing Jun 19-Jul 10], late rest [no grazing Jul 10-Jul 31], and no grazing) were replicated four times during a 13-week grazing period each summer (May 15-Aug 15). Pre (May) and post (Sept) grazing tiller (NWSG only) and plant (FORB) densities were collected each year. Grazing treatment had no effect on FORB or NWSG density. Native FORB, SG, and BBIG density was influenced ($P < 0.0001$) by date of counts with FORB density greater in May (5.13 plants/ft^2) than Sept (2.14 plants/ft^2). Our results indicate that mortality occurred during both grazing seasons, whether due to grazing pressure, animal trampling, or interspecies plant competition.

[1]Graduate Research Assistant, Center for Native Grasslands Management, Dept. of Forestry, Wildlife, and Fisheries, University of Tennessee – Knoxville, Knoxville, TN (jondrich@vols.utk.edu); Professor, Center for Native Grasslands Management, Dept. of Forestry, Wildlife, and Fisheries, University of Tennessee – Knoxville, Knoxville, TN (pkeyser@utk.edu)

IMPACT OF DEFOLIATION HEIGHT ON CROWN AND SOIL TEMPERATURE

Garrett Hatfield[1] and Chris Teutsch[2]

Defoliation management in pastures and hayfields can impact the productivity and persistence of cool-season grasses. Past research has shown that defoliation height and frequency impacts both leaf area and energy reserves in the plant. However, limited data is available on the impact of defoliation on the microclimate in cool-season grass stands. The objective of this study was to evaluate the impact of defoliation height and frequency on soil and crown temperatures of an established cool season grass stand. This study was conducted at the University of Kentucky's Research and Education Center, located near Princeton, KY. The experimental design was a randomized complete block with four replications. An established tall fescue (*Schedonorus arundinaceus* (Schreb.) Dumort., nom. cons.) stand was defoliated to 1.0 and 4.5-inches weekly and monthly to 4.5 inches. Crown and soil temperature were measured at 15 minute intervals using HOBO MX2303 loggers (Onset Computer Corporation, Bourne, MA). The 1.0-inch defoliation height increased crown temperature by more than 10 degrees Fahrenheit on the days with air temperature that exceeded 90 degrees Fahrenheit. Soil temperature was also higher at the 1-inch defoliation height, but to a lesser extent. Our data indicate that maintaining adequate residual leaf area in pastures and hayfields (>4.5-inches) could moderate the microclimate at and below the soil surface, which may in turn aid in the persistence of cool-season grass stands in the transition zone between the temperate northern and subtropical southern United States.

[1] Undergraduate, Dept. of Animal Science, Iowa State University, Ames (ghat73@iastate.edu);
[2]Associate Extension Professor, Dept. of Plant and Soil Sciences, Univ. of Kentucky, Lexington (chris.teutsch@uky.edu).

IMPACT OF FORAGE EXTENSION PROGRAMS FOCUSED ON GRAZING MANAGEMENT

Tayler Denman[1], Dennis Hancock[2], Roger Gates, Jeremy Kichler, Carole Knight, Charlotte Meeks, Steve Morgan, Lucy Ray, and Savannah Tanner[3]

Extension efforts are designed to inform people on research-backed information so they can make well-educated decisions for their operation. But do producers practice what we preach? This impact survey was designed to evaluate adoption of the practices taught at grazing management workshops. We emailed a survey to 246 previous attendees of beginner and advanced-level grazing management workshops from 2014-2019, and 42 people responded (17% response rate).

- As a result of attending a forage workshop, there was a 42% increase in the use of managed grazing practices and 38% of respondents switched from continuous to rotational grazing.
- 57% of producers indicated that what they learned at a workshop was a MAJOR HELP to increase their pasture productivity and 58% indicated that attending a workshop has helped with their farm profitability with 33% claiming their profitability increased by at least 10%.
- Comments indicate that economics and available cost share opportunities should be highlighted more throughout our workshops.

Overall feedback was very positive; however we do plan to incorporate many of the suggestions into future grazing management focused workshops.

[1]Program Coordinator, University of Georgia – Athens, GA
[2]Professor and Forage Extension Specialist, University of Georgia – Athens, GA
[3]Roger Gates(Whitfield County CEA); Jeremy Kichler(Colquitt County CEC); Carole Knight (Bulloch County CEA); Charlotte Meeks (Houston County CEA); Steve Morgan (Harris and Talbot County CEC); Lucy Ray (Morgan County CEC); Savannah Tanner (Emanuel County CEA), University of Georgia Extension *County agents are listed in alphabetical order

IMPACT OF POULTRY LITTER APPLICATION ON YIELD AND QUALITY OF ALFALFA GROWN IN MISSISSIPPI

J.B. Rushing[1], R.W. Lemus[2], J.G. Maples[3], and J.C. Lyles[4]

Alfalfa production in the Deep South has increased over the past 10 years due to cultivar development and improved management systems. Production increases have generated a need for more economical and sustainable nutrient management plans for in alfalfa, particularly in areas where poultry litter (PL) is readily available. A field trial was conducted in Newton, MS during the 2018-2019 growing seasons to compare PL and synthetic fertilizer applications. The trial consisted of a randomized split-plot arrangement with four fertilizer treatments [(1) 1 and (2) 2 tons PL/acre; (3) N-P-K based on PL analysis; (4) P-K based on PL analysis] and three cultivars [(1) 'Bulldog 505'; (2) 'Bulldog 805'; (3) 'AlfaGraze 600RR'] and replicated four times. Dry matter (DM) yield, stand persistence, and nutritive value [crude protein (CP); acid detergent fiber (ADF); neutral detergent fiber (NDF); and total digestible nutrients (TDN)] were assessed at each harvest. A total of 11 harvests were conducted across both years (six in 2018; five in 2019). No differences were observed for DM yield between cultivar (P = 0.3729) or fertilizer treatment (P = 0.1141). Also, no differences were observed for CP by cultivar (P = 0.7118) or fertilizer treatment (P = 0.4470). Both conventional cultivars had lower NDF concentrations than AlfaGraze in several harvests. For TDN, differences were observed by cultivar ($P < 0.0001$) and fertilizer treatment (P = 0.0452). Based on this research, decisions on cultivar selection and fertility management should be based on soil test analysis and input costs.

[1]Assistant Research/Extension Professor; Coastal Plain Branch Experiment Station, Newton, MS
[2]Forage Extension Specialist; Department of Plant and Soil Sciences, Mississippi State, MS
[3]Assistant Extension Professor; Department of Agricultural Economics, Mississippi State, MS
[4]Research Associate I; Coastal Plain Branch Experiment Station, Newton, MS

IMPROVING FORAGE QUALITY USING ALFALFA-GRASS BALEAGE INOCULANTS

Y. C. Newman, W. Lenzen, D. lenzen, and B. Herman [1]

Baleage of alfalfa-grass stands allows farmers to lock in high quality by harvesting forages at their peak regardless of weather and by avoiding losses associated with the making of just 'hay'. Advances in microbiology have allowed the industry to increase efficiency of their products but questions remain about the efficacy of alfalfa-specific inoculants to guarantee adequate fermentation and quality. Five treatments were evaluated to study the effects of microbial inoculants on fermentation of alfalfa-orchardgrass baleage in Wisconsin. They included four commercial inoculants and a water control. After 120 days, the pH of water-control baleage samples was higher compared to inoculated bales (5.46 vs. 4.53, respectively). Lactic acid and acetic acid production and their ratio for the different inoculant treated bales were > than 2.5, which were within the expected values for good fermentation, but no differences were found among inoculants. The water control treated bales had a low lactic-to acetic ratio. Ammonia production for all treatments was well below the threshold of 10% (as %CP) associated with lack of protein breakdown. Visual mold evaluation showed minimal presence of mold in the water-control treatment, but none was present in any of the inoculant treated bales. These results showed that, under the conditions of the study, applying inoculants seemed to guarantee good fermentation compared to no application (water control) but no differences were found among inoculants.

[1]Associate Professor/Forage Extension Specialist, Plant and Earth Science Department, Univ. of Wisconsin-River Falls, River Falls, WI; Research Assistant, Univ. of Wisconsin-River Falls, River Falls, WI; Research Assistant, Univ. of Wisconsin-River Falls, River Falls, WI; Research Assistant, Univ. of Wisconsin-River Falls, River Falls, WI

IMPROVING YOUR GRAZING?
CHANGE YOUR THINKING TO CHANGE YOUR OPERATION

Leon Atwell and Kirsten H. Robertson[1]

Are you trying to implement rotational grazing, extend your grazing season, improve your soils or just convince someone else to add lime? We must consider what is really being changed. Is it our practices, our equipment and tools, our farm or ranch? Yes, all the above...but it is really our thinking that must change first. Changing what goes on between our ears precedes changes to our operation and in our soil. In addition to shifting paradigms, we must also prepare and plan for the change in the operation. Making a plan is not something most of us want to do, but a plan is a guide for the brain to get it done. After that comes the tough challenge of implementing the plan. Every step of the path to change involves addressing mental stumbling blocks. The best practices for shifting a person's paradigm include personal connection, sharing new information and storytelling. Since lack of experience is the biggest mental challenge for most, it makes sense that surveyed producers' favorite ways to learn are through observation, participating in a grazing community, reading and workshops. This bolsters the impetus for developing both personal relationships and attending workshops to support all aspects of changing practices on your farm. Through a practical, interactive, hands-on experience, you will leave this workshop with a roadmap for changing your thinking (or helping others change their thinking) about improvements on your farm or ranch.

[1]Producer and Owner/Facilitator with Advancing Rural Prosperity, Inc. (latwell@nckcn.com), Farmer and Engineer/Manager with Greenville Soil & Water Conservation District (kirobertson@greenvillecounty.org)

INCORPORATING CRABGRASS TO IMPROVE SOUTHERN PASTURE SYSTEMS

J.J. Tucker and L.L. Baxter[1]

Bahiagrass (*Paspalum notatum*) is the primary forage for many cow-calf operations in the Deep South because it requires few inputs, but often does not meet the nutritional needs of growing cattle. Crabgrass (*Digitaria sanguinalis*) is a palatable annual grass with high nutritive value, is a prolific reseeder, and volunteers in pastures. Intentionally adding crabgrass to bahiagrass pastures may improve forage nutritive value and increase cow-calf performance. The objectives were to (1) compare establishment methods for interseeding crabgrass into bahiagrass and (2) compare performance of cow-calf pairs grazing bahiagrass pastures (BG) to bahiagrass pastures overseeded with an improved crabgrass (BGCG; Mojo®) in Alapaha, GA. A randomized complete block design with three replications was used to evaluate establishment methods (obj. 1). Treatments included variations of planting, disking, and pre-plant herbicide. After the first 4 harvests, cumulative forage yield from the plant-disc treatment was greatest (P = 0.05) and had generated almost 2,000 lbs/ac more yield than the bahiagrass control. Forty-eight cow-calf pairs were randomly assigned to eight, 6-ac paddocks where they grazed BG or BGCG pastures for 10 wks (Jul.-Sept.; obj. 2). Despite the ongoing drought, cows did not lose body condition, calves grazing BGCG gained ½ lb/day more than calves grazing BG, and each BGCG herd gained 110 lb more over this period (P = 0.04). Forage nutritive value for both evaluations are currently underway. These preliminary data show the viability of this integrated, bahiagrass-crabgrass, forage system for grazing in the region.

[1]Assistant Professor, Animal and Dairy Science Department, (jjtucker@uga.edu); Assistant Professor and Extension Forage Specialist, Crop and Soil Science Department (baxterl@uga.edu); University of Georgia Tifton Campus, Tifton, GA.

INVESTIGATING THE USE OF COVER CROPS IN EQUINE GRAZING SYSTEMS

J. L. Prigge[1], C. C. Sheaffer[2], A. L. Jaqueth[3], H. L. Lochner[4], and K. L. Martinson[5]

Cover crops have been utilized to extend the grazing season for ruminants; however, their use has not been explored within equine grazing systems. The objectives were to evaluate preference and forage nutritive value of annual rye (AR), winter rye (WR), purple top turnips (TU), daikon radishes (DR), and Berseem clover (BC) under horse grazing. Cover crops were seeded in monoculture and mixtures on Aug. 9, 2018 at St. Paul, MN. Sampling occurred on Oct. 24, 2018 to determine forage nutritive value prior to grazing. Four adult horses grazed for four hours on two consecutive days. Cover crops were visually assessed for forage removal on a 0 to 100% scale to determine preference ($P \leq 0.05$). Berseem clover in monoculture was most preferred (100% removal) followed by AR-BC mixture and AR monoculture. TU and DR were least preferred (<10% removal). Foliage and root components of TU and DR were analyzed separately. Daikon radish foliage had the greatest crude protein (34%), while DR root and AR had the lowest (≤21%). Winter rye and AR had the greatest neutral detergent fiber (≥40%), while DR and TU roots had the least (≤17%). The concentration of nonstructural carbohydrates was greatest in DR and TU roots (≥41%), and least in DR and TU foliage (≤17%). Despite a high nutritive value in each cover crop, horses would not consume DR and TU. This research is being repeated in the fall of 2019.

[1]Research Assistant, Dept. of Animal Science, University of Minnesota, St. Paul, MN (prigg028@umn.edu); [2]Professor, Dept. of Agronomy and Plant Genetics, University of Minnesota, St. Paul, MN (scheaf001@umn.edu); [3]Assistant Professor, Lake Campus-Agriculture, Wright State University, Celina, OH (aubrey.jaqueth@wright.edu); [4]Research Assistant, Dept. of Animal Science, University of Minnesota, St. Paul, MN (lochn018@umn.edu); [5]Professor, Dept. of Animal Science, University of Minnesota, St. Paul, MN (krishona@umn.edu).

LAMB GROWTH FROM NON-IRRIGATED TALL FESCUE-LEGUME PASTURES IN WESTERN OREGON

Y. Gultekin[1], S.J Filley[1], M. Smallman[1], D.B. Hannaway[2] and S. Ates[1]

Lamb production from tall fescue (TF) pastures established with either birdsfoot trefoil (TF-bft), subterranean clover (TF-sub), balansa clover (TF-bal), white clover (TF-whc) or a mixture of these legumes (TF-mix) were monitored in Corvallis, OR over two subsequent spring seasons. In 2018, all pasture combinations provided similar lamb liveweight gains (LWG, mean 177 g/d) in the first half of spring. However, lambs in TF-bal and TF-mix grew 31 to 41 g/d faster than those grazing T-sub, TF-whc and TF-bft ($P<0.05$) in the late spring period. While the available pasture mass was similar across groups, TF-bal and TF-mix had higher legume content (32 and 37% respectively) as compared to TF-sub (18%), TF-whc (20%) and TF-bft (7%) ($P<0.05$). In spring 2019, lambs that grazed TF-mix and TF-whc had higher LWG than those grazing other TF-legume pastures ($P<0.05$). The superior lamb growth rates obtained from TF-mix and TF-whc were associated with the higher legume content and pasture quality that were maintained in the late spring period. Of note was that the legume content of TF-bal was much lower in 2019 (15%) than in 2018 (32%). The present study confirmed that high legume content of pastures leads to greater lamb growth rate. A combination of annual clovers with perennial legumes in pasture mixtures ensures a higher legume content and longer persistence in dryland pastures.

[1] Graduate student, Department of Animal and Rangeland Sciences, (yunus.gultekin@oregonstate.edu); Faculty, Department of Animal and Rangeland Sciences, (shelby.filley@oregonstate.edu); Instructor and manager of the sheep unit, Department of Animal and Rangeland Sciences(mary.smallman@oregonstate.edu); Faculty, Department of Animal and Rangeland Sciences, (Serkan.ates@oregonstate.edu); [2]Faculty, Department of Crop and Soil Science, (David.hannaway@oregonstate.edu); all with Oregon State University, Corvallis OR.

MAKING HIGH QUALITY BALED SILAGE IN KENTUCKY

Jimmy C. Henning, Jeff Lehmkuhler, Brandon Sears, Corinne Belton,
Levi Berg, Don Sorrell, and April Wilhoit[1]

The ability to harvest moist forage as hay gives Kentucky producers many advantages, including timely harvest, higher forage quality, and less weathering loss over hay systems. The baleage system allows producers to utilize commonly available forage equipment (mowers, rakes, balers) rather than requiring choppers and silo structures or bags. To make high quality baleage, producers should cut at the proper stage of maturity, bale when the wilted forage is between 40 and 65% moisture content (MC), make tight bales to exclude oxygen, wrap bales the day of baling, and wrap bales with a minimum of four and ideally six to eight layers of UV-stablized, stretch wrap plastic. However, the fear of botulism from improper fermentation prevents many from adopting this beneficial technology. In addition, few lots of baled silage are tested for profiles of volatile fatty acids (VFA) even though elevated butyric acid is a good marker for possible presence of the botulinum toxin. With the assistance of county extension agents, a survey was initiated in 2017 to better understand the factors related to the pH and fermentation profile of baled silage. Moisture content explained 76% of the variation of total acid in silage in the forty-four samples collected in 2017-18. VFA production was low until MC approached 60%. Only one sample had elevated levels of butyric acid. Getting a fermentation panel of baled silage is a good predictor of poor fermentation and possible botulism toxicity. Kentucky producers could improve baled silage fermentation by making wetter silage.

[1]Extension Professor (jimmy.henning@uky.edu); Associate Extension Professor (jeff.lehmkuhler@uky.edu), Extension County Agents for Agriculture and Natural Resources (brandon.sears@uky.edu, Corinne.belton@uky.edu, levi.berg@uky.edu, dsorrell@uky.edu, april.wilhoit@uky.edu, University of Kentucky, Lexington, KY 40546.

MANAGING SUGARCANE APHID IN FORAGE SORGHUM: CULTIVAR AND APHICIDE IMPACTS

C.D. Teutsch, R.T. Villanueva, Z.J. Vilora, G.L Olson, and S.R. Smith[1]

Forage sorghum (*Sorghum bicolor* (L.) Moench) could provide a drought tolerant alternative to corn (*Zea mays* L.) for silage production in the upper Southern U.S. However, a new pest of sorghum, the sugarcane aphid (*Melanaphis sacchari* (Zehntner) (Hemiptera: Aphididae), could restrict use. Our objectives were to document of tolerance of forage sorghum cultivars to sugarcane aphid and to evaluate aphicide efficacy on these same cultivars. In May 2019, cultivars were planted at University of Kentucky's Research and Education Center at Princeton. The experimental design was a randomized complete block with split plot treatment and four replications. Sivanto (Flupyradifurone) aphicide was applied at a rate of 6 oz/A on 24-Aug-19 as a drench. In late September, plots were rated for sugarcane aphid damage on a 1 to 9 scale, with 1 being little or no damage and 9 being severe damage. Sugarcane aphid damage was less ($P < 0.02$) in plots treated with the aphicide, although the range of the difference was smaller than anticipated (5.5 versus 6.5). Cultivars differed ($P < 0.01$) in sugarcane aphid damage, with a range of 3.8 to 7.4. Although yield was higher for plots treated with aphicide ($P < 0.07$), the difference was relative small (11,922 versus 10,505 lb DM/A). In 2019, the use of aphicides had limited impact on sugarcane aphid damage and yield, but there was larger range in damage among cultivars. This indicates that the selection of cultivars with increased tolerance to the sugarcane aphid may provide a simple and cost effective approach to management in the future.

[1]Associate Extension Professor, Dept. of Plant and Soil Sciences, Univ. of Kentucky, Lexington (chris.teutsch@uky.edu); Assistant Extension Professor, Dept. of Entomology, Univ. of Kentucky, Lexington; Research Analysist, Dept. of Entomology, Univ. of Kentucky, Lexington; Agriculture Research Specialist Sr, Dept. of Plant and Soil Sciences, Univ. of Kentucky, Lexington; Extension Professor, Dept. of Plant and Soil Sciences, Univ. of Kentucky, Lexington.

MEASURING SOIL HEALTH TO MAKE NITROGEN MANAGEMENT DECISIONS

A.J. Franzluebbers[1]

Current N fertilization recommendations for fall-stockpiled tall fescue in many states of the southeastern and Mid-Atlantic regions of the US suggest 50 to 100 lb N/acre be applied in early September after clipping summer growth. This strategy seems reasonable based on agronomic principles, but does not adequately account for a potential pool of organic nitrogen that could be present in soil and suggest fall fertilizer nitrogen to be ineffective and a waste of time, money, and valuable resource, as well as a low-level, persistent threat to the environment. The issue is that soil testing for nitrogen has simply not been widely practiced, partly because of the focus on readily available inorganic forms, such as nitrate and ammonium, which are transient in soil. This presentation addresses the role of surface-soil organic nitrogen as a major contributing factor to plant-available nitrogen in forage production systems. In general, organic nitrogen must be transformed from an organic state to an inorganic state through the biologically mediated process of mineralization to be taken up by plants. Soil microorganisms are responsible for nitrogen mineralization. Although nitrogen mineralization can be measured from a soil sample in the laboratory, it is time consuming and highly dependent on methodology. An alternative is to measure soil-test biological activity during a short period of time to mimic the process. This presentation describes the role that soil-test biological activity could have on predicting nitrogen availability in fall-stockpiled tall fescue pastures. Knowing the level of soil nitrogen mineralization in a pasture could lead to more cost-effective nitrogen management decisions.

[1]Reseach Ecologist, USDA-Agricultural Research Service, NCSU Campus Box 7620, Raleigh NC 27695 (alan.franzluebbers@usda.gov).

MIXING THINGS UP WITH ALFALFA IN THE SOUTH: ALFALFA FORAGE SYSTEM MANAGEMENT STRATEGIES AND DECISION TOOLS FOR FORAGE-LIVESTOCK PRODUCERS IN THE SOUTHEAST

J.J. Tucker, Mullenix, M.K, C.G. Prevatt, J. Dubeux, E. Rios, K.B. Mason, and K. Groce

Introduction: While alfalfa was once the dominant perennial legume species used in the southern region of the US, the harsh environment and elevated insect pressure soon eliminated many productive alfalfa stands (Lacefield et al., 2009). The success and adoption of alfalfa in the North and Midwest is in part due to improved variety development for these regions, providing higher yields and improved quality potential. Although alfalfa is considered a minority crop in the region, the potential for integration into existing forage systems is high. However, there are still significant management and social barriers that must be overcome for adoption of alfalfa plantings and acreage to increase in the southern US. Legume integration in warmer-climate regions has historically been limited by 1) perceived lack of persistence, 2) lack of awareness of adapted varieties, and 3) knowledge of management requirements (Shelton et al., 2000). Newer alfalfa varieties have been developed with improved adaptation to hot humid growing conditions of the South, with dual use purposes (hay and grazing) that better fit the management opportunities for forage-livestock farmers in the region. In the last 10 years, there has been a renewed interest and an increasing amount of research and on-farm demonstration efforts surrounding the topic of integrating alfalfa into bermudagrass systems in the region. The objectives of this paper are to highlight 1) current research efforts in the Southeast on this system, and 2) regional Extension resources and decision tools available for use by stakeholders considering this system.

Current Research Efforts: A three-year evaluation of 'Bulldog 805' alfalfa interseeded into 'Tifton 85' bermudagrass was recently conducted at the University of Georgia Tifton campus to evaluate forage yield, nutritive value, and growth distribution of mixtures when harvested as baleage (Hendricks et al., in review). Seasonal yield of mixed alfalfa-bermudagrass was greater than that of Tifton 85 alone (mean 8,422 vs. 17,110 lb DM/acre) across the three-year study. This was in part due to the ability to harvest mixed stands earlier in the spring and later in the fall than bermudagrass alone. The addition of alfalfa provided up to an additional two cuttings per year in this management system (Hendricks et al., in review). Overall nutritive value of the mixed grass-legume system ranges from 60 to 63% TDN and 12 to 22% CP per cutting, with greater quality associated with increasing proportion of alfalfa in the mixture.

A multi-state alfalfa-bermudagrass clipping trial was established in fall 2017 to evaluate clipping frequency and intensity effects on mixed stands. The purpose of the study is to provide target height and rest periods for future grazing research trials, and to better understand the effects of these harvest targets on forage production and persistence of alfalfa in bermudagrass systems. Treatments included harvesting alfalfa-bermudagrass mixtures at 2-, 4-, or 6-inch

[1] Assistant Professor, University of Georgia, Tifton, GA; Associate Professor, Auburn University/Alabama Cooperative Extension System, Auburn, AL; State Specialist, UF-IFAS Range Cattle Research and Extension Center, Ona, FL; Associate Professor, UF-IFAS North Florida Research and Education Center, Marianna, FL; Assistant Professor, University of Florida, Gainesville, FL; PhD Candidate, Auburn University, Auburn, AL; and Graduate Research Assistant, University of Georgia, Tifton, GA, respectively.

clipping heights, and a 2-, 4-, or 6-inch harvest frequency. 'Bulldog 505' alfalfa was established into 'Tifton 44' and 'Russell' bermudagrass in Crossville, AL and Watkinsville, GA, respectively. In the southern half of these states, 'Bulldog 805' was planted into 'Tifton 85' bermudagrass. When mixed stands were harvested every 4 weeks to a 4-inch height, average yield was between 5,000 to 7,500 lb DM/acre for the season, which is reflective of yield potential observed for a first-year mixed stand (Hendrick et al., in review). When clipped at heights of 2-inches, alfalfa presence decreased dramatically, and less than 10% alfalfa contribution was observed in stands by mid-to-late summer. Forage mixtures harvested at a 6-inch height had decreased yield potential relative to the 4-inch height. This may be in part due to shading and competition effects of residual forage dry matter remaining post-harvest. The 6-week rest period between harvests had decreased nutritive value characteristics compared with those harvested at 2- or 4-weeks. At this harvest frequency, alfalfa had generally begun to bloom, which supports a decrease in forage nutritive value with increasing forage maturity. A second year of data collection was conducted in 2019 to further determine the effects of these harvest management strategies on persistence of alfalfa in bermudagrass systems.

Decision Tools for Stakeholders: Current research efforts and on-going on-farm demonstrations in the Southeast with alfalfa-bermudagrass systems have allowed us to update and refine our Extension recommendations for successful establishment and management of mixed systems in the region. Alfalfa requires a high-level of management and starting with the right site and conditions for establishment are critical for success. Several resources (Table 1) have been developed in the last decade to help answer stakeholder questions for those who may be considering the integration of alfalfa into their production system. Scan the QR code for each resource below to view these publications, webinars, and budget tools for on-farm decision making.

Conclusions and Next Steps: While the increase of alfalfa acreage in other regions of the US has been relatively flat in recent years (USDA NASS, 2017), there is a measurable increase in alfalfa educational efforts, plantings, and adoptions in the Southeast US in the last five years. During this time period, based on reported seed sales in Georgia, approximately 28,000 acres of alfalfa have been planted in the region (America's Alfalfa and Athens Seed Company, personal communication, 2019). As on-farm integration of alfalfa into bermudagrass systems continue to expand in the region, next-step research efforts include the evaluation of broadening the application of these systems to growers in the region. Dual-purpose use of alfalfa-bermudagrass mixtures for both conserved forage production and grazing should be considered. Many producers may harvest fields early in the spring/summer growing season, and may consider late-season grazing for mixed stands. Late-summer/early-fall stockpiling of mixed stands may also be a potential option for extending the grazing season length into the late-fall. Understanding how these management practices influence stand persistence of mixed swards, animal performance, and economics of the system will provide recommendations on additional potential applications of this forage management system in the Southeast US.

Acknowledgements: Funding for this project provided by USDA NIFA Alfalfa and Forage Research Program Grant #2017-04168

Literature Cited:

Lacefield, G.D., D.M. Ball, D. Hancock, J. Andrae, and R. Smith. 2009. Growing Alfalfa in the South. National Alfalfa and Forage Alliance.

Shelton, H., S. Franzel, and M. Peters. 2005. Adoption of tropical legume technology around the world: Analysis of success. Tropical Grasslands. 39:198-209.

Table 1. Alfalfa Management Resources

Resource	Description	QR Code
Alfalfa in Bermudagrass Checklist	A short, quick reference guide highlighting establishment considerations for alfalfa-bermudagrass mixtures	
Growing Alfalfa in the South	A regional guide on establishment, management, and economics of alfalfa production in the South	
Alfalfa Management in Georgia	An Extension publication highlighting management considerations for alfalfa systems including establishment, harvest strategies, weed control, diseases, insects, and forage utilization	
Extension Budget – Establishing and Grazing Alfalfa in Bermudagrass	An Excel spreadsheet that allows users to evaluate costs and inputs associated with alfalfa-bermudagrass mixtures	
Getting Started with Alfalfa in Bermudagrass	A webinar highlighting basic requirements for establishment of alfalfa in bermudagrass systems in the Southeast US	
Alfalfa in Bermudagrass Research, Economics and Success Stories	An overview of recent research related to alfalfa-bermudagrass mixtures in the Southeast	

NUTRITIVE VALUE OF CEREAL RYE AND RYE-TURNIP MIX IN AN INTEGRATED CROP-LIVESTOCK SYSTEM

C. Valdez Nolasco, J. L. Cox-O'Neill, D. D. Harmon, M. H. Poore, A. J. Franzluebbers, J. H. Eisemann, and C. L. Pickworth[1]

Farmers in the Southeast U.S. looking for greater sustainability of land use while land is not occupied for corn production could graze beef stocker calves on cool-season annual forages in the winter. This study evaluated forage nutritive value of either monoculture rye or 60% rye and 40% turnip mixture on two farms in eastern NC. Forage samples were collected from triplicate paddocks (6-9 acres/paddock) on both farms once every two weeks during the winter/spring grazing season (stocking rate of 1.5 acre/steer). In the rye-turnip mixture, both turnip leaf and bulb were collected. Samples were freeze dried, ground to 1 mm, and composited within forage treatment for laboratory analysis. Rye monoculture had greater organic matter (92.4%), neutral detergent fiber (46.2%), and acid detergent fiber (23.7%) than rye-turnip mixture (90.4%, 38.4%, and 20.7%, respectively). There was a tendency for greater crude protein (13.2%) in rye monoculture than in rye-turnip mixture (12.5%). Conversely, the rye monoculture had lower ash content (7.5%) than rye-turnip mixture. Based on results of low fiber content and relativity similar crude protein content with rye-turnip mixture compared to rye monoculture, this alternative cover crop mixture exhibited high forage nutritive value that could improve cattle growth and offer a unique opportunity for cattlemen in the region.

[1]Undergrad Research Assistant, NC State University, Raleigh, NC (cvaldez@ncsu.edu); Ph.D. Candidate, NC State University, Raleigh, NC (jlcox8@ncsu.edu); Extension Livestock Specialist, NC State University, Raleigh, NC (ddharmon@ncsu.edu); Ruminant Nutrition Specialist, NC State University, Raleigh, NC (mhpoore@ncsu.edu); Research Ecologist, USDA-ARS, Raleigh, NC (alan.franzluebbers@usda.gov); Professor, NC State University, Raleigh, NC (joan_eisemann@ncsu.edu); Research Associate, NC State University, Raleigh, NC (clpickwo@ncsu.edu).

NUTRITIVE VALUE OF COOL-SEASON FORAGES FOR GRAZING

Y. C. Newman, C. Snyder, and A. Neitzke [1]

Pastures are pivotal in the nutrition of grazing livestock. The amount of nutrients will be variable depending on species, maturity, nutrients in the soil or age of the stand. During early summer in the upper Midwest, many of the cool-season forage species experience a slump in production resulting in less available feed for grazing livestock and this can be compound when utilizing old stands with lack of fertility. Using a set of seven different pasture in a grazing module with low fertility in west Wisconsin, and to establish a baseline for improved management, a characterization study was initiated to (i) evaluate the nutritive value (crude protein, dry matter digestibility, total fiber, and lignin) of the pasture species present used to support grazing for dairy heifers (ii) determine the botanical composition, and (iii) measure changes in production to support the importance of pasture management. Crude protein was highest in late spring (20%) and late summer (23%), and lowest in mid-June (15%) for orchardgrass, timothy/festulolium, and bromegrass paddocks. Meadow fescue/red clover paddocks had less fluctuations, with CP averages of 18% during summer. Tall fescue had the lowest CP overall. Similar trends were found for DM digestibility for the same forages. The botanical composition rendered predominance of grasses (89%) and low legumes percent (11%). Despite the low legume population, the baseline in nutritive value should increase with improved fertility and management. Follow up evaluations are needed.

[1]Associate Professor/Forage Extension Specialist, Plant and Earth Science Department, Univ. of Wisconsin-River Falls, River Falls, WI; Research Assistant, Univ. of Wisconsin-River Falls, River Falls, WI; Research Assistant, Univ. of Wisconsin-River Falls, River Falls, WI.

OPPORTUNITIES FOR IMPROVEMENT IN TALL FESCUE MANAGEMENT

M.H. Poore[1] and S.R. Freeman[2]

Tall fescue dominates pastures in the fescue belt, which includes all or part of the mid-west, mid-south and mid-Atlantic. Most tall fescue is Kentucky-31, and is highly infected with the endophytic fungus *Epichloë coenophiala.* Fescue toxicosis includes a variety of symptoms related to vasoconstriction or depressed prolactin release. Strategies to reduce toxicosis include species diversification, introducing novel endophyte tall fescue to replace a portion of the old stand fescue, and/or adopting a strategy to mitigate toxicosis. Management approaches to mitigate fescue toxicosis include nutritional support, genetic selection of animals, and managing to minimize toxins in the plant, but no single practice has much impact on general toxicosis. Tall fescue is the premier species for stockpiling for winter grazing, and could improve the efficiency of most farms in the fescue belt. Over three years, winter grazing demonstrations were conducted on 22 farms (Freeman et al., 2019. Translational Animal Science, https://doi.org/10.1093/tas/txz086). Farmers were provided an "Amazing Grazing Kit" of temporary fence supplies (tread in posts, corner posts, reels, polywire and fence fault finder), and were supported during strip-grazing by a team of local advisors. Producers involved in the project learned how to use temporary fence, saved $28/cow, and 93% adopted the technology for use elsewhere on their farms. Novel endophyte technology has been available for 20 yrs. KY-31 has a role in many grazing systems in the fescue belt, but converting part of the pasture to novel endophyte will benefit animal performance and welfare, and will improve economic outcomes.

[1]Department Extension Leader, Department of Animal Science, North Carolina State University, Raleigh, NC (matt_poore@ncsu.edu). [2]Extension and Research Associate, Department of Animal Science, North Carolina State University, Raleigh, NC.

OPTIMAL FORAGE SPECIES SELECTION FOR SOUTHEASTERN TRANSITIONAL OR LOW-INPUT PRODUCTION

J.O.C. Kubesch, R. Nave, S. Cui, A. Griffith, G.E. Bates, and D.M. Butler[1]

Organic products are becoming popular and consumers seek natural and certified organic beef. There is limited organic forage in the southeast and the present study sought to identify forage species selections for transitioning to certified organic, natural, or low-input operations. The objective of this study was to evaluate different forage species transitioning from conventional to organic practices in Middle Tennessee. Species included tall fescue (*Schedonorus arundinaceus*) monoculture, bermudagrass (*Cynodon dactlyon*) monoculture, tall fescue and alfalfa (*Medicago sativa*) mixture, bermudagrass and alfalfa mixture, and an annual rotation mixture composed of winter wheat [*Triticum aestivum*] and winter pea [*Pisum sativum*] followed by a sorghum-sudangrass [*Sorghum bicolor x Sorghum sudanese*] and cowpea [*Vigna unguiculata*] mixture. Evaluations occurred in Spring Hill, Tennessee. Treatments were established starting in October 2017. Monthly sampling began in May 2018 for forage mass, weed competition, and nutritive value. Season and treatment interactions were significant ($P<0.05$). All treatments showed high weed presence (>20% forage mass in at least one season). Tall fescue and annual mixtures offer consistent forage production throughout the growing season. Fewer field operations involved with tall fescue and tall fescue-alfalfa are favorable to a part-time producer, though the annuals may be more productive with higher nutritive value for dairy operations.

[1]Graduate Research Assistant, University of Tennessee, Department of Plant Sciences, Knoxville, TN (jkubesch@vols.utk.edu); Assistant Professor, University of Tennessee, Department of Plant Sciences, Spring Hill, TN (rnave@utk.edu); Associate Professor, Middle Tennessee State University, Department of Agriculture, Murfreesboro, TN (Song.Cui@mtsu.edu); Associate Professor, University of Tennessee, Department of Agricultural and Resource Economics (agriff14@utk.edu); Professor, University of Tennessee, Department of Plant Sciences, Knoxville, TN (gbates@utk.edu); Associate Professor, University of Tennessee, Department of Plant Sciences, Knoxville, TN (dbutle14@utk.edu).

OPTIMAL FORAGE SPECIES SELECTION FOR SOUTHEASTERN TRANSITIONAL OR LOW-INPUT PRODUCTION

J.O.C. Kubesch, R. Nave, S. Cui, A. Griffith, G.E. Bates, and D.M. Butler[1]

Abstract

Organic products are becoming increasingly popular with American consumers. In farmer's markets consumers generally see a variety of fruits and vegetables; however, consumers are looking for natural and certified organic beef. While there is a body of organic research in the northern United States and Europe, there is limited research in the southeast. Organic forage production is a small portion of the southeastern market. The present study seeks to identify the best forage species selections for producers transitioning to certified organic production.

In addition to certified organic production, the present work also supports farms seeking to produce natural beef or low-input settings. The research study examined the following forage mixtures and monocultures in Middle Tennessee: tall fescue (*Schedonorus arundinaceus*) monoculture, bermudagrass (*Cynodon dactylon*) monoculture, tall fescue and alfalfa (*Medicago sativa*) mixture, bermudagrass and alfalfa mixture, and an annual rotation composed of winter wheat [*Triticum aestivum*] and winter pea [*Pisum sativum*] mixture rotated with sorghum-sudangrass [*Sorghum bicolor x Sorghum sudanese*] and cowpea [*Vigna unguiculata*] mixture.

The study was conducted at the University of Tennessee, Middle Tennessee AgResearch and Education Center in Spring Hill, Tennessee. Treatments were established from October 2017 to October 2018. The tall fescue and bermudagrass monocultures received 75 lb N/ac from horse manure in the spring of 2019. Monthly sampling began in May 2018. Forage mass, weed competition, and nutritive value were measured over the course of the growing season.

Establishment issues occurred in the bermudagrass plots as well as alfalfa. Bermudagrass and bermudagrass-alfalfa treatments showed high weed competition (up to ~90% of forage mass was weed). Managing existing weed problems is a major challenge to transitional systems. Overall, the tall fescue monoculture and the annual forage rotation offered the most consistent forage base for systems transitioning to organic production in the southeast U.S. The reduced number of field operations involved with tall fescue and tall fescue-alfalfa stands appears to make these

[1] Graduate Research Assistant, University of Tennessee, Department of Plant Sciences, Knoxville, TN (jkubesch@vols.utk.edu); Assistant Professor, University of Tennessee, Department of Plant Sciences, Spring Hill, TN (rnave@utk.edu); Associate Professor, Middle Tennessee State University, Department of Agriculture, Murfreesboro, TN (Song.Cui@mtsu.edu); Associate Professor, University of Tennessee, Department of Agricultural and Resource Economics (agriff14@utk.edu); Director, University of Tennessee, Beef and Forage Center, Knoxville, TN (gbates@utk.edu); Associate Professor, University of Tennessee, Department of Plant Sciences, Knoxville, TN (dbutle14@utk.edu).

selections favorable to a part-time producer, even though the annual forage rotation appears to be more productive.

Introduction: Rapid urbanization in Middle Tennessee, specifically the Nashville metropolitan area, promote a growing demand for organic products. Farmers markets and farm-to-table interactions generally focus on vegetable production and value-added goods. However, these consumers also seek organic-labelled beef and dairy options. Cattle operations in this region are small, generally composed of 35 to 40 animals on average. The pastures in these farms are composed of either tall fescue or bermudagrass. Also, these operations likely use low-input production practices, such as limited herbicide, fertilization, or liming (Neil Rhodes, personal communication; Fribourg and Waller, 1997). Most producers in the region focus on beef cow-calf operations. While there has been a decline in dairy producers, remaining operations seek high-quality forage to overcome animal performance issues in the region.

Tall fescue is a cool-season perennial grass that is widely used in the southeast grasslands (Hoveland, 2009). The persistence and agronomic simplicity of the species drove adoption in the 1940s and 1950s. Management to reduce endophyte side-effects has improved animal performance (Hoveland et al., 1980; Hoveland et al., 1983; Roberts et al 2009). In order to maintain consistent forage availability across the growing season in the southeast, especially during summer, warm-season perennial grasses such as bermudagrass are used to complement tall fescue pastures (Ball et al., 2015). In addition, alfalfa being a cool-season perennial legume, may also complement tall fescue and bermudagrass production (Hoveland, 1986).

Winter and summer annuals are similarly used to meet high nutritive value requirements for livestock and to meet forage needs when permanent pasture is insufficient. These forages are popular in dairying and stockering operations (Ball et al., 2015; Hoveland, 1986). Winter wheat (*Triticum aestivum*) and Austrian winter pea (*Pisum sativum*) are popular cover crops in grain operations. Sorghum-sudangrass (*Sorghum bicolor x Sorghum sudanese*; [sudex]) and cowpea (*Vigna unguiculata*) are used as forages in the summer and early fall.

Transitioning to certified organic production can be expensive and laborious for producers. However, developing a clear pathway to transition with a low-input forage program can potentially increase transition success rates. The present study seeks to identify the ideal forage species selection for transitional organic production. The hypothesis was that cool-season perennial mixtures would provide consistent forage mass of high nutritive value throughout the growing season as compared to other forage species.

Materials and Methods: Plots were established at the University of Tennessee, Middle Tennessee AgResearch and Education Center in Spring Hill, TN (35°43'07.2"N 86°57'54.8"W), as a randomized complete block design with 5 species treatments and 4 replications. Previously, the site was composed of an orchard which had received no synthetic fertilizer since 2016. The original ground cover included the following: bermudagrass, tall fescue, white clover (*Trifolium repens*) and Virginia pepperweed (*Lepedium densiflorum*).

Plots were established in October 2017 and measured 75 ft^2. Tall fescue (c.v. Kentucky 31), seeded at 20 lb/ac, and winter annuals (wheat c.v. LG 334 SRW; winter pea variety not stated)—wheat seeded at 100 lb/ac and winter pea seeded at 50 lb/ac—were drilled into conventional seedbeds October 27, 2017. Bermudagrass (variety not stated), seeded at 6 lb/ac, was hand-sown June 4, 2018. Alfalfa (c.v. WL 358 LH), seeded at 15 lb/ac, was drilled March 16, 2018. Summer annuals (sudex c.v. AS 6501; cowpea c.v. Iron and Clay blend)—sudex seeded at 30 lb/ac and cowpea seeded at 75 lb/ac—were conventionally drilled June 4, 2018. Bermudagrass (c.v. common) failed to emerge in the spring planting, so it was reseeded on June 27, 2018 at a rate of 15 lb/ac for monocultures and 10 lb/ac for the bermudagrass and alfalfa mixture Wheat and winter pea were reestablished on October 10, 2018 following the same protocol stated above. Plots were harvested monthly from May to November 2018 with a walk-behind silage chopper to a 5" stubble in order to maintain an actively growing stand and control weeds. On August 24, 2018, all plots were cut down to 3" prior sampling throughout the growing season. Winter annuals were planted in October 10, 2018. Bermudagrass (c.v. Cheyenne II) was re-established in June 2019 by the same rates and methods in order to improve stand as well as rectify a seed selection issue. Alfalfa failed to establish in 2018 (<1 plant/ 1-ft^2), therefore, replanting was conducted May 14, 2019 and again October 8, 2019. Plots with alfalfa received 1.78 lb/ac B March 7, 2019. Tall fescue monocultures received 75 lb N/ac by means of horse manure March 7, 2019; bermudagrass monocultures received the same N rate from the same manure source March 7, 2019. Plots with alfalfa received boron at 1.78 lb/ac on March 7, 2019. Re-establishment occurred to ensure that treatment comparisons would be between swards with target plant populations. Summer annuals were planted again by the same method and rates June 5, 2019. Winter annuals were planted again by the same method and rates October 8, 2019.

Monthly samples were collected from October 2018 to September 2019. Harvesting prior to October 2018 was not included for production and conducted only for establishment purposes. Two 1-ft^2 samples were cut down to 2-in from each plot and then separated by species to determine forage and weed percentages for botanical composition. Forage mass and botanical composition were determined following drying the components for 72 hours at 136°F. Following forage sampling, plots (with exception of summer annuals) were harvested down to a 3" stubble height with a silage chopper to prevent rank growth. Summer annuals were harvested down to 6" monthly.

Nutritive value [crude protein (CP) and neutral detergent fiber (NDF)] was determined based on methods described by Murray and Cowe (2004). Forage and weed components were recombined within each sample and ground through a Wiley Mill with a 1-mm screen (Model 4). Samples were subsequently homogenized with a cyclone sample mill (UDY Corporation, Fort Collins, CO). Samples were scanned using the 2018 Grass Hay equation from the NIRS Forage and Feed Consortium (NIRSC, Hillsboro, WI). Win ISI II software was used for NIRS analysis (Infrasoft International, State College, PA). The Global H statistical test compared the samples against the model and samples from distinct data sets within the database for accurate results (Murray and Cowe, 2004).

The study was analyzed in SAS as a mixed model ANOVA using the DandA macro. (SAS v 9.4; Saxton, 2013). Monthly data were pooled into seasons. September, October, and November were pooled as Fall; December, January, February were pooled as Winter; March, April, May were pooled as Spring; and June, July, August were pooled as Summer. Mean separation was done by Tukey's Honest Significant Differences at an alpha of 0.05.

Results and Discussion: Species by season interactions were significant for forage mass across the year (Table 1; $P<0.01$). The annual rotation mixture produced the highest forage mass during summer (Table 1). The tall fescue monoculture and tall fescue-alfalfa mixture had similar forage mass, consistently across months. The bermudagrass and bermudagrass-alfalfa forage mass was variable throughout all seasons ($P<0.05$). Alfalfa re-established in May 2019 failed during summer, with fewer than 5 plants/ft^2 by August. Alfalfa establishment issues have been reported in similar organic systems, which may be due to *Sclerotinia* crown rot or boron deficiencies during establishment (Eichler-Inwood et al., 2015).

Bermudagrass and bermudagrass-alfalfa mixture resulted in greater weed competition than all other treatments through all seasons (Table 2; $P<0.01$). Weed competition across species was lowest in the winter months. Crabgrass (*Digitaria* spp.) was the predominant weed grown across all plots, and since it is a warm-season annual grass, its growth pattern is similar to bermudagrass plants. In addition, crabgrass can also be consumed by livestock and in many instances, it is considered a forage with high nutritional value, which confounded weed identification and management. Noxious weeds were also present, including palmer and spiny pigweeds (*Amaranthus* spp.), which Steckel (2007) suggests will remain a challenge in the southeast. Sanderson et al. (2012) reported that grass-clover mixtures can better compete against weeds than monocultures, though this observation was not confirmed in the present study. Bermudagrass sprigs are usually chemically treated, therefore not allowed for establishment under organic systems. Due to reduced available non-treated seeded bermudagrass varieties and given timeline of the current study, re-establishment period came at expense of production and favored additional weed establishment.

Crude protein ranged from 8% to 23% within and across seasons and treatments (Table 3; $P<0.01$). The annual rotation, tall fescue monoculture, and tall fescue-alfalfa mixture had most of the time higher CP than the bermudagrass and bermudagrass-alfalfa treatments. For the annual treatment, CP was lowest in the summer and highest in the winter (Table 3). For the tall fescue and tall fescue-alfalfa treatments, there were no seasonal differences in CP (Table 3). These values fall within the target range presented by Ball et al. (2015).

Neutral detergent fiber was lowest for the annual treatment across all seasons except summer (Table 4; $P<0.01$ for fall, spring winter; $P=0.28$ for summer). Tall fescue and tall fescue-alfalfa treatments had lower NDF content than the bermudagrass monoculture and bermudagrass-alfalfa mixture (Table 4). Overall nutritive value was greatest in the annual rotation in the annual rotation, tall fescue, and tall fescue-alfalfa.

Balancing forage quantity and quality over the growing season has been a major challenge in organic systems (Eichler-Inwood et al., 2015). Tall fescue and annual selections were more consistent at producing high-quality forage over the season than the bermudagrass mixture and monoculture. Based on data from 2019 production, the optimal species selection appears to be either an annual rotation or a tall fescue monoculture or mixture. Crop establishment and weed competition are two challenges identified in the first production year.

Literature Cited:

Ball, D.M., C.S. Hoveland, and G.D. Lacefield. 2015. Southern Forages (5th Edition). International Plant Nutrition Institute (IPNI), Peachtree Corners, GA.

Eichler-Inwood, S.E., G.E. Bates, and D.M Butler. 2015. Forage performance and soil quality in forage systems under organic management in the southeastern United States. Agron. J. 107(5):1641-1652.

Fribourg, H.A., and J.C. Waller. 1997. Native and other warm season perennial grasses in Tennessee. Research Report 97-12. Agricultural Experiment Station. Knoxville, TN.

Hoveland, C.S. 1986. Beef-Forage Systems for the Southeastern United States. J. An. Sci. 63(3): 978–985.

Hoveland, C.S. 2009. Origin and history. p. 3–10. In H.A. Fribourg, D.B. Hannaway, and C.P. West (ed.) Tall fescue for the twenty-first century. Agron. Monogr. 53. ASA, CSSA, and SSSA, Madison, WI.

Hoveland, C.S., R. L. Haaland, C. C. King, Jr., W. B. Anthoily, E. M. Clark, J. A. McGuire, L. A. Smith, H. W. Grimes, and J. L. Holliman 1980. Association of *Epichloe typhina* fungus and steer performance on tall fescue pasture. Agron. J. 72: 1064-1065.

Hoveland, C. S., S. P. Schmidt, C. C. King, Jr., J. W. Odom, E. M. Clark, J. A. McGuire, L. A. Smith, H. W. Grimes, and J. L. Holliman. 1983. Steer performance and association of *Acremonium coenophialum* fungal endophyte on tall fescue pasture. 75: 821-824.

Murray, I., and I. Cowe. 2004. Sample preparation. Near infrared spectroscopy in agriculture 1: 75-112.

Roberts, C.A., G.D. Lacefield, D. Ball, and G.E. Bates. 2009. Management to optimize grazing performance in the northern hemisphere. p. 85–100. In H.A. Fribourg, D.B. Hannaway, and C.P. West (ed.) Tall fescue for the twenty-first century. Agron. Monogr. 53. ASA, CSSA, and SSSA, Madison, WI.

Sanderson, M.A., G. Brink, L. Ruth, and R. Stout. 2012. Grass-legume mixtures suppress weeds during establishment better than monocultures. Agron. J. 104:36–42.

Saxton, A.M. 2013. DandA.sas: Design and analysis macro collection version 2.11. University of Tennessee, Knoxville, TN.

Steckel, L.E. 2007. The Dioecious *Amaranthus* spp.: Here to Stay. Weed Tech. 21:567-570.

Table 1. Average forage mass across seasons of different forage species and mixtures produced under a low-input organic system in Tennessee during 2019.

Treatments	*P*-value across season	Forage Mass Fall		Spring		Summer		Winter	
		--lb/ac--							
Annual	<0.01	2170	a,C	3830	a,B	10475	a,A	633	a,D
Bermudagrass	<0.01	2419	a,AB	1512	b,B	3575	bc,A	966	a,B
Bermudagrass-Alfalfa	<0.01	2766	a,A	1246	b,B	4860	bc,A	711	a,C
Tall Fescue	>0.05	1730	a,A	2820	a,A	1770	c,A	1803	a,A
Tall Fescue-Alfalfa	>0.05	1917	a,A	2776	a,A	1835	c,A	1613	a,A
P-value within season		0.12		<0.01		<0.01		0.08	

Treatments with the same lowercase letter do not differ within season. Treatments with the same uppercase letter do not differ across seasons. $P = 0.05$.

Table 2. Weeds competition across seasons of different forage species and mixtures produced under a low-input organic system in Tennessee during 2019.

		Weeds Competition							
Treatments	*P*-value across season	Fall		Spring		Summer		Winter	
		--- % ---							
Annual	<0.01	21	c,A	3	b,B	9	c,A	3	b,B
Bermudagrass	<0.01	88	a,A	65	a,AB	82	a,AB	55	a,B
Bermudagrass-Alfalfa	>0.05	91	a,A	76	a,A	90	a,A	39	a,B
Tall Fescue	>0.05	57	b,A	7	b,B	45	b,A	5	b,B
Tall Fescue-Alfalfa	>0.05	67	ab,A	15	b,B	67	ab,A	9	b,B
P-value within season		<0.01		<0.01		<0.01		<0.01	

Treatments with the same lowercase letter do not differ within season. Treatments with the same uppercase letter do not differ across seasons. $P = 0.05$.

Table 3. Average crude protein across seasons of different forage species and mixtures produced under a low-input organic system in Tennessee during 2019.

Treatments	*P*-value across season	Crude Protein							
		Fall		Spring		Summer		Winter	
		------------------------------------ % ------------------------------------							
Annual	<0.01	17	a,B	19	a,B	11	ab,C	23	a,A
Bermudagrass	<0.01	10	c,AB	12	b,A	7	b,B	8	d,B
Bermudagrass-Alfalfa	<0.01	11	bc,AB	14	b,A	9	ab,B	10	cd,B
Tall Fescue	>0.05	14	ab,A	12	b,A	13	a,A	12	bc,A
Tall Fescue-Alfalfa	>0.05	13	abc,A	12	b,A	14	a,A	13	b,A
P-value within season		<0.01		<0.01		<0.01		<0.01	

Treatments with the same lowercase letter do not differ within season. Treatments with the same uppercase letter do not differ across seasons. P=0.05.

Table 4. Average neutral detergent fiber across seasons of different forage species and mixtures produced under a low-input organic system in Tennessee during 2019.

Treatments	*P*-value across season	NDF							
		Fall		Spring		Summer		Winter	
		---%							
Annual	<0.01	51	b,A	40	c,B	56	a,A	26	d,C
Bermudagrass	<0.01	60	a,B	48	ab,C	60	a,B	73	a,A
Bermudagrass-Alfalfa	<0.01	60	a,A	45	bc,B	60	a,A	65	b,A
Tall Fescue	<0.01	54	ab,AB	54	a,AB	60	a,A	49	c,B
Tall Fescue-Alfalfa	>0.05	51	b,A	54	a,A	56	a,A	48	c,A
P-value within season		<0.01		<0.01		0.28		<0.01	

Treatments with the same lowercase letter do not differ within season. Treatments with the same uppercase letter do not differ across seasons. $P = 0.05$.

OPTIMAL NITROGEN USE ON SUMMER ANNUAL FORAGE MIXTURES

K.M. Mercier, C.D. Teutsch, S.R. Smith, E.L. Ritchey, K.H. Burdine, and E.S. Vanzant[1]

Legume inclusion into perennial pastures has traditionally been linked with increased productivity coupled with reduced need for nitrogen fertilizer. However, limited information exits as to whether these same phenomena are realized in annual pastures. A study was conducted in 2018 and 2019 at two locations in Kentucky (Princeton, Zanesville silt loam with a fragipan; Lexington, Maury silt loam) to determine the effects of nitrogen application rate (0, 50, 100, 150, and 200 lb N/A) on three different sorghum-sudangrass based forage treatments (monoculture, simple mixture-3 species, and complex mixture-12 species; both mixtures contained legumes). Plots were harvested three times each year at 30-40 in. In three out of the four location+year combinations (environments), forage treatment had no effect on yield ($p>0.06$). However, for three out of the four environments, N treatment still had a linear effect on yield ($p<0.01$), even though legumes were included in mixtures. These results indicate that increasing biodiversity in summer annual pastures may not offer any additional yield benefit, as well as the need to still apply nitrogen fertilizer on these pastures even with legume inclusion.

[1]Graduate Student, University of Kentucky Research and Education Center at Princeton (UKREC), Princeton, KY (kelly.mercier@uky.edu); Extension Associate Professor, UKREC, Princeton, KY (chris.teutsch@uky.edu); Extension Professor, University of Kentucky (UK), Lexington, KY (raysmith1@uky.edu); Extension Associate Professor, UKREC, Princeton, KY (elritc2@uky.edu); Associate Extension Professor, UK, Lexington, KY (kburdine@uky.edu); Associate Professor, UK, Lexington, KY (evanzant@uky.edu).

OUTREACH ON GRAZING LANDS TO ENHANCE ECONOMIC ANALYSIS (COST BENEFITS) FOR CONSERVATION CHANGES

Statement of current relevance or need: The National Grazing Lands Coalition and our project partner (The Matrix Assessment Group, LLC) are conducting a 3-year outreach/education/demonstration project on how a variety of management and conservation practices impact pasture and range productivity, economics, and sustainability. The project is investigating the role economics plays on producer decisions about conservation program participation, on the design of grazing management systems and to what extent multispecies management influences economic performance. To this end, data have been collected from a large group of producers located in several different regions of the U.S.

As the project enters its second-year additional data are being collected through an online system. Producers are being asked to assemble economic data that links improvements due to conservation practices to economic returns. Very little information is available reflecting how flexible herd management for rangeland and pastureland operations is linked to soil health, stocking rates, drought response and economic performance. Data from a large sample size of producers will produce results that can be used both for communication to other producers and for linking these results to soil health indicators. This linkage will allow us to develop a very powerful campaign on the benefits of soil health on all types of grazing lands operations. We anticipate that tying economic information to conservation practices will have a significant impact on financial incentives and influence the adoption of different management/conservation practices in the future.

Producers are sharing how they evaluate and manage soil health, water quality, plant and animal resources, and productive capacity on their grazing lands. Combining these responses with a variety of socio-economic information will facilitate a larger scale economic analysis.

The development of the Secretary of Agriculture's Climate Hub-based vulnerability studies has provided some broad scale indicators of how soil health will be affected by management decisions. The addition of an economic analysis is the last piece of the puzzle that producers ultimately need to make management decisions that lead to significant improvements to their operations. Analyses eventually will include spatial analyses to see if there are economic differences due to location, size of operation or other selected factors.

General methods and data analysis information: The project database currently contains information from over 80 producers from six regions of the U.S. Workshops have been conducted in Oklahoma, Texas, South Carolina, Florida, Pennsylvania and New Hampshire as these states have active grazing land coalitions that worked to organize and host workshops. In several cases producers from neighboring states participated in the regional workshops. Participating producers varied considerably in each region, including part vs. full-time, small scale to larger ranching operations and family vs. corporate operations. Data are from new producers as well as those representing multi-generational operations.

Some data will be summarized in tabular/graphic form. However, much of the information collected is in text format and will be analyzed using some unique text data analytic procedures.

Results or predicted results: Thus far the project includes responses from 82 producers with a total of 80,376 acres. Respondents indicate on average they manage just less than two thirds of all those acres with conservation in mind. Operations include ones as small as 5 acres and as large as 12,000 acres with the average sized operation consisting of 980 acres.

The poster will highlight some of the results that might be of most interest to those attending the SWCS conference. For example, all respondents indicated that soil health and quality are very important to their operations. A summary of the frequency with which they employ selected practices to assess and manage soil health and quality will be presented. A similar summary for water quality and quantity also will be shared.

One of the most critical sections of the survey includes an inventory of management and conservation practices being employed by producers. These include those provided with assistance from NRCS associates both with and without funding attached. It also includes those practices in which producers are engaged without outside assistance or funding.

Respondents prioritized a list of issues that concerned them as they manage their operations several of which are related to soil and water issues. These responses lend themselves to graphical analysis and will be part of the poster presentation.

Explanation of how the results or outcomes contribute to science and society: This project is innovative because it: 1) targets producer-led grazing groups that include underserved conservation audiences; 2) engages the targeted audience in independent, on-going conservation management decisions that will improve their soil's health and thus the land's sustainability for economic and environmental objectives; 3) develops a large scale economic & spatial analysis of grazing management practices; and 4) develops economic information in key areas associated with grazing lands.

PASTURE SCHOOL FOR CENTRAL FLORIDA LIVESTOCK PRODUCERS

Bainum, C.[1], Hersom, M., Wallau, M., Silveira, M., Bosques, J., Strickland, J.S., Justesen, B., Mann, M., Yarborough, J.K., Jennings, E., Cooper, C., Walter, J., Gamble, S., Rivera, F., Mudge, D., and Wilson, T.

Recognizing the unique advantage an environment conducive to year-round grazing provides to livestock producers in the state of Florida can greatly affect the bottom line of their operation. This two-day intensive pasture school was developed by the Central Florida Livestock Agent's Group in an effort to improve pasture management for utilization in livestock operations. The learning objectives were to gain a knowledge increase on the topics of soil health, forage establishment and management with regard to warm and cool-season pasture, weed control, grazing strategies, and matching cattle nutrient requirements with available forage. Topics were delivered by UF/IFAS Livestock Extension Agents and State Extension Specialists. Each day included a combination of classroom lectures and hands-on experiential learning opportunities such as herbicide spray rig calibration, analyzing soil composition changes, calculating herbage availability in the field, and touring research plots of various Florida forages. This program had 34 attendees, representing over 20,000 acres of pasture land. Pre and post tests were administered which suggested a 52% knowledge gain on soil health topics, an 82% increase in knowledge of grazing management, and a 72% increase in knowledge related to cattle nutrition. This program provided livestock producers with fundamental training on forage basics to enhance all areas of their production, centered on a 365-day pasture base.

[1]Extension Agent, UF/IFAS, Marion County, FL (cbainum@ufl.edu)

PERFORMANCE AND FEED QUALITY EVALUATION OF SELECTED FORAGE SPECIES IN A MIXED CROPPING AGROFORESTRY SYSTEM IN DESERT ENVIRONMENT

M. A. Islam[1], M. K. Suleiman[1], A. M. Quoreshi[1], M. Jallow[1], and S. Abbas[1]

Abstract

The effects of climate change and soil degradation are expected to significantly reduce native plant cover in coming years to put more strain on the availability of livestock feed. Despite having a thriving demand, forage production has been very limited in Kuwait. Currently, 95% livestock feed are imported, and it is the main constraint to sustainable livestock production. Considering the fact that cultivation of forage species in mixed cropping system can increase on-farm biodiversity, ameliorate soil and microclimate, as well as ensure the year-long production of feeds, we evaluated the performance and nutrition quality of Leucaena (*Leucaena leucocephala*), a leguminous perennial, smooth bromegrass (*Bromus inermis* L.), Sorghum (*Sorghum bicolor*), and Spring Triticale (x Triticosecale). Leucaena and sorghum exhibited good growth. However, smooth bromegrass and spring triticale did not grow very well due to gradual increase in ambient temperature, suggesting that these species should be grown in Kuwait preferably between November and early April. Nutrition analysis revealed 13.7%, 15.9%, 18.2% and 20.8% crude protein in sorghum, triticale, bromegrass, and Leucaena, respectively. These results suggest that the high protein content of these species make them ideal to be grown locally as forage or feed mix to reduce dependency of importing feed from abroad.

Introduction: Agriculture is limited in Kuwait due to lack of water and arable land. Besides, available cultivable land is often degraded due to persistent harsh environmental conditions, anthropogenic disturbances, and uncontrolled animal grazing (Brown, 2003). These factors have put huge pressure on livestock owners to produce or import additional feed for their livestock. Only 5% of the total feed requirement (dry matter basis) is produced in Kuwait as fresh alfalfa/grass and grain by-products and rest (95%) is imported. The high cost of imported feed is a major constraint to sustainable production of livestock in Kuwait. Although there is a thriving demand for feeds, forage production in Kuwait has not been very diverse.

Monoculture of crops, as currently being practiced by farmers in Kuwait to produce livestock feed, is often focused on maximizing productivity of a single crop, and this practice is associated with high input of fertilizers and pesticides (Malezieux et al., 2009). In contrast, mixed species cropping is viewed as a viable tool to increase on-farm biodiversity in agriculture production systems and is a potentially important component of any sustainable cropping system (Ramert et al., 2002). Other benefits can also include improved soil structure, improved weed control and microclimate amelioration (wind barrier) (Vandermeer, 1989).

---

[1]Research Scientist, Desert Agriculture and Ecosystems Program, Kuwait Institute for Scientific Research, P.O. Box 24885 Safat 13109, Kuwait.

Considering the aforementioned factors and the year-long demand for livestock feed, we studied the performance and feed quality of *Leucaena leucocephala*, a leguminous tree species, and three grass or crop species: smooth bromegrass (*Bromus inermis* L.), sorghum (*Sorghum bicolor)*, and Spring Triticale (x Triticosecale) in single and mixed cropping systems. It is expected that mixed cropping cultivation of *Leucaena, bromegrass*, sorghum, and Spring Triticale will ensure year-long supply of forage as livestock feed, and also contribute towards amelioration of soil properties through increased soil microbial population diversity and biological nitrogen fixation. Successful implementation of this production system will reduce the pressure of uncontrolled grazing on native plant populations, thereby, help protect the desert ecosystem, support natural regeneration of native plants, and negate the potential effects of climate change.

Materials and Methods: The experiment was conducted in House of Development Company for Agricultural Contracting farm (N 29° 22' 99.3", E 47° 80' 30.8") at Sulaibiya, Kuwait. The research area and the surrounding areas exhibited typical extreme arid region characteristics. For example, the annual mean air temperature of the area was 25.9°C. Mean highest monthly temperature (48°C) was observed in August, while January experienced the coldest mean temperature (8°C). Generally, cooler months between December and March experience occasional small amount of rainfall, hence, considered growing season for most native plants in Kuwait. Annual average rainfall is meagre, around 113 mm, while evapotranspiration can reach up to 4000 mm (Almedeij 2012). Texture of the soil was loamy sand and soil pH ranged from 7.8 to 8.3. Average soil bulk density was 1.44 g/cm^3 and ECe of the soil from the experimental plots ranged between 0.74 mS/cm to 4.55 mS/cm. The average soil organic matter content was 1.09%.

Before commencing the study, the land was cleared, debris were removed, and soils were mixed and loosened up by using a Rotavetor and levelled. Water tank, pump, and sprinkler irrigation system were installed. Seed viability of sorghum, spring triticale, and Leucaena was determined through the 2, 3, 5-triphenyl tetrazolium chloride (TTC) method and for smooth bromegrass by the germination method. Seed viability ranged from 76% to 100%. *Leucaena leucocephala* seeds were germinated in the laboratory and later transferred to the greenhouse for hardening. Seedlings of Leucaena and seeds of sorghum, triticale, and smooth bromegrass were planted or sown during the last week of March 2019. Seedlings of *Leucaena leucocephala* were planted at 2 m x 2 m spacing in pure and mixed-cropping plots. Seeds of sorghum, Spring Triticale, and smooth bromegrass were sown in strips that were 30 cm apart in both pure and mixed-cropping plots. The seeding rate of 22.45 kg/ha for bromegrass, 7 kg/ha for sorghum, and 112 kg/ha for spring triticale was adopted following conventional growing practices for the aforementioned species for forage production (Bean, 2017; Agriculture Victoria, 1996).

Each plot size was 100 m^2 and the plots were laid out in randomized complete block design (RCBD) with six treatments consisting of a control treatment, Leucaena 100%, smooth bromegrass 100%, sorghum 100%, triticale 100%, and Leucaena, sorghum, triticale and smooth bromegrass in rows/strips (mixed cropping), each replicated three times. One hundred kilogram of organic fertilizer and 32 kg of urea were applied to the soil prior to planting/seed sowing.

Plants were irrigated daily based on the evapotranspiration loss of water for crops like sorghum growing in desert conditions (Allen et al. 1998). Despite regular irrigation, symptoms of sun-scorching effect on leaves were evident by necrotic tips. Therefore, sorghum, triticale, and bromegrass were harvested ahead of schedule in late June 2019.

Results and Discussion: Data were analyzed using one way Analysis of Variance (ANOVA). Means were separated using Duncan's multiple range analysis. All transplanted *Leucaena leucocephala* seedlings survived and exhibited good growth. Overall, by end of June 2019, there was about two and half fold increase in height and number of branches of the transplanted seedlings while root collar diameter increased by 132%. Dry matter yield of Sorghum, smooth bromegrass, and triticale in mono cropping was 1868 kg/ha, 682 kg/ha, and 656 kg/ha, respectively. In mixed cropping total dry matter yield was 800 kg/ha. The biomass yield for these crops were relatively low compared to the published studies for the arid regions. This is due to the sharp increase in the ambient temperature from early March to June in Kuwait. Therefore, it would be prudent to grow these crops as early as mid-November or early December to maximize their growing period and yield. *Leucaena leucocephala* leaves were not harvested as harvesting was planned when the seedlings will be one-year old. Nutritional analysis of these crops revealed 13.7%, 15.9%, 18.2% and 20.8% crude protein in sorghum, triticale, bromegrass, and Leucaena, respectively (Table 1). These results suggest that the high protein content of these species make them ideal to be grown locally in the desert conditions as forage or feed mix.

Literature Cited:

Agriculture Victoria. 1996. Growing Triticale. Note Number: AG0497. The State of Victoria, Australia. Available at http://agriculture.vic.gov.au/agriculture/grains-and-other-crops/crop-production/growing-triticale, Accessed on July 03, 2019.

Allen, R. G., L. S. Pereira, D. Raes, and M. Smith. 1998. Crop evapotranspiration-Guidelines for computing crop water requirements. FAO Irrigation and Drainage Paper No. 56. Food and Agriculture Organization, Rome, Italy.

Brown, G. 2003. Factors maintaining plant diversity in degraded areas of northern Kuwait. J Arid Environ. 54: 183–194.

Malezieux, E., Y. Crozat, C. Dupraz, M. Laurans, D. Makowski, H. Ozier-Lafontaine, B. Rapidel, S. De Tourdonnet, and M. Valantin-Morison. 2009. Mixing plant species in cropping system: Concepts, tools and models. A review. Agron Sustain Dev. 29 (1): 43–62.

Ramert, B., M. Lennartsson, and G. Davies. 2002. The use of mixed species cropping to manage pests and diseases–theory and practice. p. 207–210. In Proceedings of the UK Organic Research 2002 Conference, Organic Centre Wales, Institute of Rural Studies, University of Wales, Aberystwyth.

Vandermeer, J. H. 1989. The Ecology of Intercropping. Cambridge, UK: Cambridge University Press.

Table 1. Nutrient concentration in selected forage species. Different letters after the standard error of means indicate significant differences at p ≤0.05.

Species	Parameters (% ± SEM)			
	Ash	Crude Fat	Crude Fiber	Crude Protein
Sorghum	14.50 ± 0.26 b	1.29 ± 0.10 c	22.19 ± 0.36 a	13.75 ± 0.54 b
Triticale	28.05 ± 6.67 ab	1.56 ± 0.42 bc	18.00 ± 1.98 a	15.98 ± 1.65 ab
Bromegrass	29.31 ± 4.66 a	3.13 ± 0.41 b	18.31 ± 1.76 a	18.23 ± 2.48 ab
Leucaena	15.18 ± 1.27 b	8.62 ± 0.79 a	13.10 ± 0.40 b	20.80 ± 0.25 a

PHENOTYPIC VARIATION OF FORAGE TYPE BERMUDAGRASSES

Robert A. Lane, Chukwudi Nwaogwugwu, and Shyam Nair[1]

An earlier study confirmed the genetic uniqueness of a selection of a dense, fast-growing bermudagrass (cv. "Raven") that appears to be a natural hybrid (no seed production detected). While initial observations indicate Raven is at least moderately rust resistant, this study was designed to examine phenological differences between Raven and other hybrids in widespread use (Jiggs and Tifton-85) as well as other apparent natural hybrids from the same general vicinity in Huntsville, Texas. Dry matter production during the study did not differ significantly among cultivars. Leaf length, leaf width, and stolon length of Raven and Jiggs were virtually identical. Tifton-85 produced significantly faster vertical growth and longer leaves than all others. While Jiggs produced longer stolon internodes, Raven produced a similar stolon growth rate and significantly more stolons than T-85 and Jiggs. No significant differences were detected in rooting speed or depth of rooting from excised stolons in a controlled greenhouse study. Stolon production, stolon growth rate and fast rooting all play an important role in speed and ease of establishment of asexually (sprigged) improved bermudagrasses. Further studies of Raven and other hybrids are underway to evaluate other physical traits, including rhizome production and cold tolerance. Forage quality characteristics, stability of rust resistance, preferred soil adaptation, clipping tolerance, and variable rate N fertilization studies are being planned as well. Locally adapted hybrids may be superior to commercially available cultivars.

[1]Professor (blane@shsu.edu), Graduate Research Assistant (cun003@shsu.edu), and Associate Professor (shyam.nair@shsu.edu), respectively, Department of Agricultural Sciences, Sam Houston State University.

POTASSIUM FERTILIZATION AND ITS IMPACT ON YIELD, QUALITY, AND WINTER HARDINESS OF ALFALFA

A. Greenberg, M.T. Berti, D. Franzen, A. Peterson, S. Cabello, A. Wittenberg, and K. Mozea[1]

Alfalfa (*Medicago sativa* L.) is the most widely grown forage crop in the world. An important nutrient for improving alfalfa's yield, quality, and winter hardiness is potassium. There are multiple factors that should determine the appropriate rate of potassium fertilization, one being the soil's clay mineralogy. When the soil's smectite-to-illite clay ratio is greater than 3.5, potassium becomes immobile within the soil and is less accessible to the crop. The experiment was conducted at Milnor and Lisbon, ND, in 2019, both fields had low K in the soil (<100 mg/kg K) and two different smectite-to-illite ratio (high and low). Three alfalfa varieties with fall dormancy scores of 3, 4, and 5 were fertilized with five different treatments of potassium. The treatments were 0, 168, and 336 kg K_2O/ha at seeding, and 168 and 336 kg K_2O/ha as split-application. The FD3 variety had a lower yield, which was expected. There was a slight response to yield with increased kg K_2O/ha (r^2=0.62). This study formed a stronger understanding of the most appropriate potassium rates for alfalfa to maximize quality, yield, and winter survival. By better understanding the most efficient rates of potassium application, depending on the soil's clay mineralogy, growers can more economically, produce higher quality and yield of alfalfa.

[1] Graduate Research Assistant (amy.greenberg@ndsu.edu); Forage and Biomass Production, Professor (marisol.berti@ndsu.edu); Soil Science, Professor (david.franzen@ndsu.edu), Research Specialist (alan.peterson.2@ndsu.edu); PhD student (sergiofabian.cabello@ndsu.edu); MS student (alex.1.wittenberg@ndsu.edu); MS student (kenneth.mozea@ndsu.edu) North Dakota State University, Fargo, ND.

RED CLOVER DOSED IN LOOSE MINERAL MITIGATES VASOCONSTRICTION IN STEERS CHALLENGED WITH E+ TALL FESCUE SEED

B.E. Harlow[1], J.T. Johns[2], and M.D. Flythe[3]

Endophyte-infected (E+) tall fescue can cause persistent vasoconstriction in ruminants and consequently fescue toxicosis. Legumes, including red clover (RC), contain isoflavones that are known hypotensive agents. Previous experiments have shown the benefits of RC (pasture or hay at 15 – 30% of the diet) on blood flow and growth performance of cattle grazing E+ tall fescue, but the minimum level of RC supplementation to elicit vascular benefits has not been identified. The objective of this study was to determine if low levels of RC in conventional loose mineral could mitigate vasoconstriction associated with E+ tall fescue seed challenge. Rumen fistulated steers received 1 of 4 treatments: control mineral (6 oz head d^{-1}), 10% RC mineral, 15% RC mineral, or 20% RC mineral. All steers were first adapted to a high fiber diet for 2 weeks *ad libitum* and then were subjected to a 2-week E+ seed challenge. The caudal artery of each steer was imaged repeatedly using Doppler ultrasonography to monitor artery luminal areas. The E+ seed challenge induced vasoconstriction in control steers receiving conventional loose mineral. Both 10% and 15% RC mineral treatments did not reliably mitigate vasoconstriction associated with E+ seed challenge. However, 20% RC mineral steers remained dilated and were unaffected by the E+ seed challenge. These results indicate that 20% RC mineral could mitigate vasoconstriction associated with tall fescue toxicosis. Future field research is needed to evaluate the efficacy of 20% RC mineral in cattle grazing E+ tall fescue pastures.

[1]Post Doctoral Fellow, USDA-ARS, Forage-Animal Production Research Unit, Lexington, Kentucky (brittany.harlow@usda.gov); [2]Beef Specialist, Burkmann Feeds, Danville, Kentucky (jjohns@burkmann.com); [3]Supervisory Research Microbiologist, USDA-ARS, Forage-Animal Production Research Unit, Lexington, Kentucky (michael.flythe@usda.gov).

REVIEW OF THE OREGON PRODUCER DISCUSSION GROUPS

Woody Lane[1]

Since 1995, three Forage Discussion Groups in western Oregon have conducted more than 640 monthly meetings that include pasture walks, technical discussions, and guest speakers. These private, dues-supported groups of grass-based farms and ranches meet monthly on host operations, led by a facilitator. The main features and conclusions from this program are:

- General model: Producer groups can be financially self-supporting. Each group has its own bank account, treasurer, and paid facilitator. Annual dues are $200 per year. This model works.
- Meetings: Frequent meetings are important, and members will attend them. Meetings include a pasture walk and an indoor technical discussion, sometimes with a guest presenter.
- Information: Members learn about advanced topics and discuss the latest developments.
- Diversity: Members raise many types of livestock, many perspectives. They also include seed companies, agricultural supply stores, veterinarians, university faculty, and agency personnel.
- Facilitator: A good facilitator is critical for the ongoing success of a producer group.
- Closed membership: Only members (and guests) can attend meetings. This allows honest, in-depth discussions, especially about finances & goals. An underlying rule is confidentiality.
- Prerequisite: No beginners; members must have knowledge from courses or experience.
- Making changes: A very effective way of transferring technology is through peer-to-peer interactions. We watch what good producers do. Groups provide connections in a structured way.
- Reinforcement: Member comments & evaluations reinforce new practices and reduce the risks of making changes. Producer groups provide a support network for trying new ideas.

[1]Livestock Nutritionist, Lane Livestock Services, Roseburg, OR (woody@woodylane.com).

REVIEW OF THE OREGON PRODUCER DISCUSSION GROUPS – MONEY IN THE BANK

Woody Lane[1]

Introduction: I am going to review a program that has been incredibly valuable and successful for so many livestock farmers and ranchers. For 25 years here in Oregon, we've quietly been running producer discussion groups, every month, month after month. Members have gained immense benefits from these groups. Some producers have been involved since the groups began. And these groups have always supported themselves financially and haven't cost the public a dime. So, it's about time we review some of the main lessons from these groups. We can even call it our quarter-century review.

Firstly, some terminology. A few of these groups exist around the country, and they are known by many names: producer groups, grazing groups, grazing networks, study groups, forage groups. (I don't believe anyone has called them "producer flocks", "producer herds", or "producer pods".) But let's move past the semantics. At their core, these are organized or semi-organized groups of producers who periodically gather and share ideas. Meetings may be monthly, bimonthly, four times a year . . . whatever. These groups may or may not include a formal facilitator. The members may or may not pay dues, and the dues may vary from trivial to substantial. As you'd expect, there's lots of variation.

But I'll focus on the model I'm most familiar with — the three groups that I facilitate here in western Oregon: The Umpqua Valley Forage Study Group (UVFSG) began in 1995. The Forage And Nutrition Group (FANG) on the South Coast began in 2000. And the Willamette Valley Grazing And Nutrition Group (WVGANG) began in 2002.

So let's explore, summarize, and review.

Producer groups work: We'll start with the main lesson: *these groups work*. The model we developed in Oregon provides superb value to the members. These groups help member farms stay in business during difficult times and thrive in times of good prices. And the groups are financially self-sustaining.

Finances: *Money talks*. Producers are willing to pay good money to be members. Our dues are $200 per year, with $100 paid every six months. Each group has its own bank account, arranges for a treasurer (a member of the group), and pays the facilitator who is essentially a contractor working for the group. These dues fly in the face of history because American agriculture has a long tradition of providing information for free or very cheaply. (For example, many full-day Extension meetings may cost $15 which includes lunch. Find *that* in other industries.)

[1]Livestock Nutritionist and Forage Specialist, Lane Livestock Services, Roseburg, OR (woody@woodylane.com).

Sustainability: This level of dues also allows these groups to be self-supporting and sustainable. And because these groups don't rely on public funds, either from grants or as salaries for public employees, they are not affected by the vagaries of government and university budgets and priorities.

Meetings: Having frequent meetings is important, and members will attend them. These groups meet monthly or nearly monthly. Meetings usually last three hours. Meetings are held on member operations and usually include a pasture walk and then an indoor technical discussion, occasionally with guest presenters. As of this writing, we've held a total of 643 meetings since 1995. You read that right: 643.

Information: Every month, we engage in detailed discussions about advanced topics like stocking density, Phase II versus Phase III growth, mob grazing, soil health, cation exchange capacity, buffer index, the grazing wedge, K-Line irrigation, soil test grids, RFQ, arbuscular mycorrhizae. Also new forages like the hybrid forage brassicas, plantain, balansa clover, BMR millet, fodder beets, improved crabgrass, grazing brome, Italian ryegrass planted in the spring. Members become very knowledgeable about new products and technologies. Essentially, members are professional livestock operators engaged in discussing the latest developments in their field.

Diversity among members: Diversity is important, especially in experience and ideas. These groups are not beef cattle groups or dairy groups or young farmer groups (or old farmer groups). The members raise many types of livestock – sheep, beef cattle, dairy cattle, bison, hogs, pastured poultry, horses. Membership also includes seed companies, agricultural supply stores, veterinarians, university professors, and agency personnel (NRCS, SWCD). Members vary in age from their late 20s to well into the 70s. Consider for a moment the vast range of experiences and philosophies all these folks bring to the table. These differences are a valuable strength. But having a facilitator to monitor and umpire is a practical way of dealing with strong opinions and personalities.

Facilitator: As the facilitator of these three groups, I am not exactly unbiased. But I have also conducted surveys of other producer groups around the U.S. and have been in contact with equivalent groups in New Zealand. One unmistakable conclusion: A good facilitator is critical for the ongoing success and sustainability of a producer group.

Closed membership: These groups have specific membership lists, although these change over time, of course. Only members (and their guests) can attend the meetings. Meetings are not open to the public and are not advertised. People know each other, and with that knowledge comes a certain level of trust. This allows members to talk honestly about sensitive issues like finances, labor, goals, and personal doubts. An underlying rule of these meetings is confidentiality. Like the old New Zealand saying about woolsheds, "What's said in the shed, stays in the shed."

Acronyms: Good acronyms are cool. They are easier to remember, and they roll of the tongue with pleasure. And they're cool.

Prerequisite: Early in our history, members decided to establish a prerequisite for new members: namely, they must have attended formal courses in forages or grazing (like my courses) or have lots of good experience as knowledgeable graziers. No beginners. To an outsider, this may seem elitist, but it's actually critical to success. A prerequisite keeps the discussions on a high level so that meetings don't focus on the beginner issues that usually dominate public meetings. This helps retain some of the most experienced and successful producers in the groups.

Prerequisite is not a new concept. Universities use it every day for their classes. For example, graduate student courses are always off-limits to freshman and sophomores. In agriculture, we're not used to prerequisites, but this criterion has worked extremely well for these producer groups.

Sharing experiences: In every meeting, there are *hundreds* of years of experience in the room. You can't buy advice like that. As one member once said, "With all this experience in the room, I can't live long enough to make all the mistakes that these folks have made, so I should learn from them now." Well said.

Making changes: Research has shown that one of the most effective ways of transferring technology is through peer-to-peer interactions. Translation: we watch what the neighbors do. And in these producer groups, the "neighbors" are the other group members whom you know and trust. Month after month, you hear comments and suggestions that are constructive and insightful. So you feel that you make changes with confidence and support. And we have observed over the years that group members have indeed made many changes in their operations.

Reinforcement: With most management decisions on a farm, the concept of *risk* is the 800-pound gorilla in the room. We don't usually talk about it or include it in our spreadsheets, but privately we stay awake at night thinking about it. But risk is really a main factor in making decisions and trying out new ideas. These groups help reduce risk. They provide a support network for members who embark on a new venture or try a new technique or product. Every meeting gives everyone opportunities to ask for help, receive evaluations, receive good advice about options and adjustments. These conversations have prevented many "wrecks." And it's always good to hear comments from people you trust.

Each meeting is exciting and valuable. Every month we meet and talk, and sometimes the conversations veer into unusual territories that surprise and delight everyone. Periodically, we also have guests who give presentations, either in person or by online video technology. As the facilitator, I set the agenda for each meeting, but that agenda is really just a suggestion. The discussions lead where they will go, and that's where the members want them to go.

But in the end, the discussions are serious, and they focus on economics. Not necessarily the micro-economics of a particular pasture or a piece of equipment, but on the business economics of the *entire* operation, on how all these pieces fit together to support a profitable enterprise. And everyone in the room helps explore the ramifications and nuances of that economics. That's the bottom line.

These producer groups work extremely well.

How do I know these groups work? Here's one way: the annual dues are $200 per year. Many members have been attending for at least 10 years, sometimes more than 20 years. Here's a question for you: would *you* voluntarily make out a check for $100 every six months for something that you didn't value?

I'll end by paraphrasing the comments of one astute member who is a very good livestock producer and businessman. He said, "What's important is not how much you produce or how big your farm is. It's how much money is in the bank at the end of the day."

REVIEWING THE FIRST DECADE OF BERMUDAGRASS STEM MAGGOT RESEARCH

Lisa Baxter, Bill Anderson, Dennis Hancock, Will Hudson, and Chris Prevatt[1]

Since it was discovered in South Georgia in the summer of 2010, the bermudagrass stem maggot (BSM; *Atherigona reversura* Villenueve) has severely damaged bermudagrass (*Cynodon dactylon*) pastures and hayfields throughout the Southeastern US. Although the degree of damage depends on the bermudagrass variety, latitude of the farm, and time of year, producers have reported up to an 80% yield loss in late summer. The economic impact of the BSM damage depends on several factors, but if a conservative yield loss of 25% is generally applied to just the bermudagrass acreage in Georgia, the potential economic loss totals $30 million. Georgia researchers have conducted over a dozen research projects since the BSM first appeared a decade ago. These include screening insecticides, refining spray recommendations, selecting new varieties with improved tolerance, and understanding the basic life cycle of the pest. Our research has already answered several important questions but it is critical that work continues to address the lack of diversity with our BSM management tools. Preventing overuse and uneconomical use of insecticides is a crucial educational objective to combat the potential for resistance. Funding for this work was provided by the Georgia Beef Commission.

[1]Assistant Professor and Extension Forage Specialist, Crop and Soil Science Dept., University of Georgia, Tifton, GA; (baxterl@uga.edu); Research Geneticist, USDA-ARS, Tifton, GA (bill.anderson@ars.usda.gov); Forage Extension Specialist, Crop and Soils Science Dept., University of Georgia, Athens, GA, (dhancock@uga.edu); Extension Specialist, Entomology Dept., University of Georgia, Athens, GA (wghudson@uga.edu); State Specialized Extension Agent II – Beef Cattle and Forage Enterprise Budgeting and Marketing, Range Cattle Research and Education Center, Ona, FL (prevacg@ufl.edu).

RUMINANT AGRICULTURE: ESSENTIAL FOR SUSTAINABLE SOCIETIES

Adele Hite, PhD, Eric Westman, MD, Peter Ballerstedt, PhD[1]

The American Forage and Grassland Council promotes the role of forage agriculture in sustainable food systems, which are essential to sustainable societies. For four decades, the United States Department of Agriculture (later joined by the Department of Health and Human Services) has advocated diets with limited red meat and full-fat dairy product consumption – the primary products of forage agriculture. The primary rationale for this policy was and is that such diets will reduce the incidence of chronic, non-communicable diseases, but environmental issues have also been a consideration. The evidence base for the dietary patterns recommended has remained controversial since their inception. The incidence of chronic diseases has increased despite the American public changing their eating habits. A large and growing body of research and clinical results strongly suggests that a new approach to dietary advice is needed. The dietary patterns advocated by the DGA are not intended for the treatment of disease. A majority of adult Americans are metabolically ill to some degree. Public health messages suggest that obesity is a cause of (or a risk factor for) chronic diseases, while research demonstrates that it is a symptom of metabolic illness. Once more, a large and growing body of research data and clinical experience has demonstrated the role of an appropriate dietary intervention in reversing metabolic illness. That diet, however, is fundamentally different from the "healthy diet" promoted by public agencies and private interests. The economic, societal and environmental burdens of this epidemic is unsustainable. The productivity and efficiency of our global ruminant agriculture systems must be improved if Sustainable Development Goals are going to be met.

[1]Adele Hite, PhD, North Carolina State University; Eric Westman, MD, Duke University Medical School; Peter Ballerstedt, PhD, Barenbrug USA, pballerstedt@barusa.com

SAFE AND TASTY SORGHUM

Shelby Gruss, Mitch Tuinstra, and Keith Johnson [1]

Sorghum is an important forage crop. All current sorghum varieties produce a cyanogenic glycoside called dhurrin that accumulates in high concentrations in young tillers or plants stressed by frost or extreme drought. Dhurrin is degraded to produce hydrogen cyanide that can cause toxicity to grazing animals. In this study, a new dhurrin-free sorghum hybrid developed through mutagenesis was tested against current market hybrids for preferential feeding by sheep. The three market hybrids selected were Sweet 6 BMR Dry Stalk (bmr-6 sorghum x sudangrass), Sweet Bites (sorghum x sudangrass), and Greentreat® Rocket (bmr-6 sudangrass x sudangrass). Biomass was sampled before and after grazing, an unmanned aerial vehicle was flown each day to document the amounts grazed over time, and trail cameras were used to track sheep movement throughout the plots. Based on the grazing data, the dhurrin-free hybrid was preferentially grazed with approximately 51% of the forage grazed, followed by Sweet 6 BMR Dry Stalk ~28% grazed; percentages are based on dry biomass. Sheep spent the most time in Sweet 6 BMR Dry Stalk during cycle one and the other three hybrids were comparable in time. In the second cycle, more time was spent in the dhurrin-free hybrid in comparison to the other three hybrids. Taken together, the dhurrin-free hybrid was a good grazing option as it has an acceptable yield, is favorably fed upon, and is safe for livestock to graze young plants or stressed tissue.

[1] PhD Candidate in Plant Breeding and Genetics, sgruss@purdue.edu; Professor of Plant Breeding and Genetics, mtuinstr@purdue.edu; Professor of Agronomy, johnsonk@purdue.edu; Purdue University, Department of Agronomy, West Lafayette IN

SELECTION FOR PERSISTENCE IN RED CLOVER (*Trifolium Pratense* L.) THROUGH IMPROVED TOLERANCE TO NORTHERN ANTHRACNOSE

M. Sahin[1], S. Reid[2], E. Sahin[1], S. Ates[3], S. Elias[1], D.B. Hannaway[1]

Red clover is one of the most important forage legumes, used extensively on European and North American farms as a short-lived perennial. This study was conducted with 23 red clover genotypes to evaluate tolerance of northern anthracnose (NA) and yield in Boyd, KY and for agromorphological characteristics in Philomath, OR. Cultivars and accessions were evaluated for NA tolerance of naturally occurring inoculum using phenotypical selection. Plants were rated for disease tolerance on a 1 to 9 scale, with 1 being necrotic (susceptible) and 9 being no visible symptoms (tolerant). Dry matter (DM) production was determined for each cutting in 2018 and 2019. Differences ($P<0.05$) existed for disease severity and yield. In 2018, 'Evolve' and 'Redkin' had highest DM yields (5078 and 4917 kg/ha, respectively), while in 2019, 'Helike' and 'Evolve' had the highest DM yields (9923 and 9843 kg/ha, respectively). NA disease tolerance ranged from 2.6 to 7.8, with 'Kenland' being most tolerant. Agromorphological characteristics were determined in 2019 for the same 23 red clover genotypes grown under irrigation in Oregon. Differences were observed for plant height, spring growth, and growth habit, indicating potential for developing cultivars suitable for both grazing and hay/silage systems. Future work will include rating space plant nursery for another selection cycle.

[1] Graduate student, Department of Crop and Soil Sciences, OSU, Corvallis, OR (muhammet.sahin@oregonstate.edu); [2]Chief Breeder and Research Director, USA DLF Pickseed USA, Corvallis, OR (sreid@dlfna.com), [1]Graduate student, Department of Crop and Soil Sciences, OSU, Corvallis, OR (elif.sahin@oregonstate.edu); [3]Faculty Department of Crop and Soil Sciences, Oregon State University, Corvallis, OR (Serkan.ates@oregonstate.edu); [1]Faculty, Department of Crop and Soil Science, OSU, Corvallis, OR (Sabry.Elias@oregonstate.edu); [1]Faculty, Department of Crop and Soil Sciences, OSU, Corvallis, OR (David.hannaway@oregonstate.edu).

SOIL HEALTH RELATIONS WITH GREENHOUSE GAS EMISSIONS IN SEMI-ARID PASTURES

K.L. Radicke, L. Slaughter, C.P. West, B.J. Petermann, D.D. Henry[1]

With increasing climate variability, effective management of soils determines its ability to perform functions and processes effectively and sustainably. Grasslands with livestock grazing rather than intensive row cropping, could help build soil health by reducing soil disturbance and providing continuous plant cover (Bhandari et al., 2018), further, the inclusion of higher quality legumes that are interseeded into these systems could potentially further enhance soil health and carbon sequestration. A field study was conducted in summer 2019. Soil gas flux measurements were taken at 0, 15, and 30 minute intervals using permanent static chambers in 3 grass-alfalfa pastures at various alfalfa densities (0-25%, 25-50%, 50-75%, 75-100%). Methane (CH_4) fluxes were not consistent in 2019 with relation to different alfalfa density; however, preliminary 2018 data and data from June 25, 2019 indicate substantially less CH_4 consumption into the soil (-2 grams/acre) compared to 2018 (-81 grams/acre). Volumetric water content in 2018 was about 23% more than in 2019. This suggests that methane consumption was likely favored by soil moisture rather than legume density. Soil microorganisms can consume or emit CH_4, therefore, managing these systems by leaving enough residual cover post grazing that enhances soil moisture content can potentially optimize soil uptake of CH_4 by providing an opportunity to reduce and buffer the climate impact of grazing ruminant livestock CH_4 emissions.

[1] Graduate assistant, Department of Plant and Soil Science, Texas Tech University, Lubbock, TX 79404; Assistant Professor, Soil Microbial Ecology, Texas Tech University, Lubbock, TX 79404; Professor, Department of Plant and Soil Science, Texas Tech University, Lubbock, TX 79404; Graduate assistant, Department of Plant and Soil Science, Texas Tech University, Lubbock, TX 79404; Assistant Professor, Sustainable Livestock Grazing Management, Texas Tech University, Lubbock, TX 79404-2122

SORGHUM-SUDANGRASS SOWN WITH COWPEA AND BALED AT TWO MOISTURE LEVELS FOR GREATER BALEAGE NUTRITIVE VALUE

C.C. Nieman[a], D. Philipp[1], K.P. Coffey[1], and J.G. Franco[2]

Sorghum-sudangrass is high-yielding, but have low crude protein and digestibility. Intercropping with an annual legume, such as cowpea, may improve nutritive value by providing highly digestible, high protein forage, while potentially increasing dry matter yields. To better preserve nutritive value compared to hay and to reduce delays in drying time, baleage is an option. We investigated the productivity and quality of sorghum-sudangrass grown in monoculture (SS) and in binary intercrop with cowpea (SS-C) when harvested and stored as baleage at two different moistures (high and low). All 1.7-acre plots were cut when sorghum-sudangrass reached 48 in. height in the afternoon of August 8 and were baled between 11:30 am and 2:00 pm on August 9, 2018. The high and low moisture treatments were 61.2 and 48.0%, respectively. Yields did not differ between SS and SS-C and averaged 2.2 tons/acre. Legume contribution was low in mixtures, ranging from 0 to 14% and averaged 2.6% of total dry matter. Cowpea plants were unable to compete with the rapid growing sorghum-sudangrass, thus did not influence nutritive value. However, moisture level affected most fermentation variables including pH, total acids %, ammonia-N (% of total N), lactic acid %, and acetic acid %, but did not affect propionic acid %, butyric acid %, and isobutyric acid %. In summary, intercropping with cowpea did not provide a nutritive value advantage due to competition from sorghum-sudangrass; however, moisture had large effects on fermentation, with higher moisture resulting in lower pH and greater volatile fatty acid production.

[1]University of Arkansas, Department of Animal Science, Fayetteville, AR (nieman@uark.edu, dphilipp@uark.edu; kcoffey@uark.edu); [2]USDA-ARS, Dale Bumpers Small Farms Research Center, Booneville AR (jose.franco@usda.gov) .

SOUTHERN WOMEN IN AG: EMPOWERING WOMEN CATTLE PRODUCERS

Knight, C.H.[1], Butcher, S.R.[2], Cheely, T.W.[3], Hammond, K.[4], Ray, L. [5], Sapp, P.[6], Tucker, J.J. [7]

Of the 17,779 women that identified as farming operators in Georgia in the U.S. Census for Ag, only 36% identified as the principle operator. It is not from lack of skill that women are not more prevalent in the industry, but perhaps lack of confidence. Increasing the confidence of women in ag settings by encouraging them to experience basic ag techniques/skills in a stress-free, all-female environment will result in greater involvement. The Southern Women in Ag (SWAG) Advanced Cattle Workshop was developed to provide women involved or interested in cattle production a comfortable learning environment to gain hands-on experience and network with other women involved in the industry. A two-day hands-on training was held April 29-30, 2019 on the UGA-Tifton Campus, Tifton, GA; 18 attendees and 3 UGA ANR county agents attended, not including volunteers and instructors. Each day consisted of three two-hour breakout sessions which allowed attendees ample time to participate in each of the hands-on activities provided. Sessions included: cattle handling/chute side, truck and trailer driving, tractor and equipment, media training, bovine reproduction, and forages and fencing. Based on the evaluation, comfort level increased by at least 1 score in every station. As a result of this program, 5,882 acres and 1,222 head of cattle will be impacted by the knowledge gained. Additionally, all attendees received BQA Certification and a one-year membership in the Georgia Cattlewomen's Association.

[1]ANR Agent, UGA Ext, Statesboro, GA (clh@uga.edu); [2]ANR Agent, UGA Ext, Coweta County, Newnan, GA; [3]ANR Agent, UGA Ext Warren County, Warrenton, GA; [4]NW Research Center Supt, UGA CAES, Calhoun, GA; [5]ANR Agent, UGA Ext Morgan County, Madison, GA; [6]ANR Agent, UGA Ext Jefferson County, Louisville, GA; [7]Assistant Prof, UGA Animal and Dairy Science, Tifton, GA.

SPRING SEEDING OF COOL-SEASON ANNUALS INTO ORCHARDGRASS IN AN EFFORT TO INCREASE FORAGE MASS

Y.V. Xiong, R. L. G. Nave, G. E. Bates, and M. D. Corbin[1]

Establishing cool-season annual grasses in spring after orchardgrass pasture stands have been diminished by drought the previous fall should improve forage production in the southeastern U.S. Annual ryegrass and forage oats were introduced into orchardgrass plots under 5 treatments in a randomized complete block design with three replicates during this 2-year research study. In 2017, after a warm and wet spring, there were no significant differences in yield between orchardgrass and either overseeded annual forage crop. However, crude protein content of annual ryegrass was significantly greater than forage oat, and NDF was significantly lower than pure orchardgrass. In 2018, forage oat produced 1700 lb more DM/acre than orchardgrass and 700 lb more DM/acre than ryegrass, but there were no significant differences in forage quality among the treatments. Burndown treatment increased the average forage production and forage quality compared to no burndown in both 2017 and 2018. This study showed that fall drought can have a long-term impact on orchardgrass pasture even when followed by a wet spring. Overseeding annual forage is a feasible method to increase forage biomass production and nutritive value.

[1]Post Doc Research Associate, UTIA, Plant Sciences, Spring Hill, TN (yxiong9@utk.edu); Assist. Prof., UTIA, Plant Science, Spring Hill, TN (rnave@utk.edu); Professor and Director of the UT Beef and Forage Center, UTIA, Plant Science, Knoxville, TN (gbates@utk.edu); Research Associate, UTIA, Plant Sciences, Knoxville, TN (mcorbin1@utk.edu).

SUBSEQUENT CORN YIELD AND SOIL STRUCTURE FOLLOWING GRAZING OF ANNUAL FORAGES IN AN INTEGRATED CROP-LIVESTOCK SYSTEM

J. L. Cox-O'Neill, C. Valdez Nolasco, D. D. Harmon, M. H. Poore, A. J. Franzluebbers, J. H. Eisemann, and C. L. Pickworth[1]

Grain producers have increased their use of cover crops for agronomic purposes, but typically don't incorporate grazing due to perceived concerns about soil compaction and negative effects on crop yields. This study evaluated the subsequent corn yield and soil structure following winter grazing of (1) rye monoculture and (2) 60% rye and 40% turnip mixture cool-season cover crop treatments on two farms in eastern NC. Three replicate paddocks per treatment at each location were grazed by stocker calves (605 lb; stocking rate of 1.5 acre/steer). A split plot of ~ 1 acre was sectioned off in each paddock to prevent grazing. Corn yield and soil samples were collected after grazing from both grazed and non-grazed areas. Corn yield at 15% moisture was not different between cover crop treatments or between grazed and non-grazed areas (108 bu/acre). Soil bulk density was not different between treatments or between grazed and non-grazed areas (1.15 g/cm^3) at a depth of 0-5 cm, but had a tendency to be greater for all of the grazed (1.47 g/cm^3) than non-grazed areas (1.41 g/cm^3) at 5-15 cm depth. Early results from the study indicate that grazing either forage treatment over the winter may have minimal negative impact on either subsequent corn grain yield or soil compaction at this stocking rate.

[1]Ph.D. Candidate, NC State University, Raleigh, NC (jlcox8@ncsu.edu); Undergraduate Research Assistant, NC State University, Raleigh, NC (cvaldez@ncsu.edu); Extension Livestock Specialist, NC State University, Raleigh, NC (ddharmon@ncsu.edu); Ruminant Nutrition Specialist, NC State University, Raleigh, NC (mhpoore@ncsu.edu); Research Ecologist, USDA-ARS, Raleigh, NC (alan.franzluebbers@usda.gov); Professor, NC State University, Raleigh, NC (joan_eisemann@ncsu.edu); Research Associate, NC State University, Raleigh, NC (clpickwo@ncsu.edu).

SUMMER COVER CROPS AND EFFECTS ON WINTER PASTURE PRODUCTION AND SOIL PHYSICAL, CHEMICAL, AND BIOLOGICAL PROPERTIES

J. K. Rogers, S.G. Robertson, G.R. Sweeten, and J.T. Biermacher[1]

Southern plains land used for wheat production is typically summer-fallowed chemically or with tillage to conserve moisture. The fallow period is an opportunity to incorporate cover crops that can reduce soil and wind erosion, add soil organic material, and increase soil microbial activity. Physical, chemical, and biological impacts of summer cover crops on the soil health, production, and economics of a wheat winter pasture system grazed with stocker cattle need to be quantified. We established a split-plot study with tillage or no-tillage as the main plot and cover crop or no cover crop as the split-plot. Each treatment is five acres with five replications. In the split-plots, a warm-season annual (50% broadleaf and 50% grass) cover crop mix is planted in May and terminated in early September. Following cover crop termination, wheat is established. Both cover crops and wheat are grazed with stocker cattle and animal performance measures are collected. Soil moisture and temperature at three depths is collected in each paddock as well as soil bulk density, water infiltration, soil microbial community, standard soil test, and Haney test. Forage mass and nutritive values is collected from both the cover crops and wheat. Within tillage treatment, wheat winter pasture production following summer cover crops is generally reduced. Grazing days of the summer cover crop have been less than 50 which helps offset but not cover the cost of establishment. Soil erosion control is improved with cover crops. Economic returns for winter pasture have been higher in treatments with no cover crops.

Associate Professor, Pasture and Range, (jkrogers@noble.org); Senior Research Associate, (sgrobertson@noble.org); Research Associate 2 (grsweeten@noble.org); Associate Professor, Applied Economics Research, (jtbiermacher@noble.org); Noble Research Institute, Ardmore, OK.

TECHNIQUES FOR EVALUATING HEALTH AND PRODUCTIVITY OF COOL SEASON PERENNIAL PASTURES

K. Lea, S. Beidleman, K. Hargadon, and S.R. Smith[1]

The UK Horse Pasture Evaluation Program has serviced the equine industry since 2005 with the goal of increasing available forage and reducing tall fescue toxicity on horse farms. The current sampling technique uses 4 ft^2 quadrat (~20 quadrats per pasture) and requires extensive time and labor to measure detailed botanical composition, tall fescue endophyte infection and ergovaline concentration. Larger farms have expressed concerns that the current technique is too time consuming, too costly, and yields more information than they are able to process. Therefore, The Pasture Health Scorecard was developed as an alternative. It was modeled after the NRCS Pasture Condition Score to quickly evaluate equine pastures in ten categories: desirable grass cover, desirable grass diversity, legumes, bare soil and warm season annual grasses, broadleaf weeds, perennial weedy grasses, grazing management, stage of growth, loafing areas and thatch/color. For each category, pastures are given a score of one to five. Farms receive a score card for each pasture evaluated, a summary for the farm, and a list of pastures "needing improvement", "improvements would be beneficial", or "no improvements needed" for each category. In 2019, Pin Oak Stud, located in Versailles, KY, agreed to test the Pasture Health Score Card. Sampling time and labor was reduced by more than half, while additional sampling indicated that recommendations from the Pasture Health Score Card were similar to those based on current quadrat sampling techniques.

[1] Research Analyst, Univ. of KY, Lexington, KY (krista.lea1@uky.edu); Technician, Univ. of KY, Lexington (sydneybeidleman@uky.edu), KY Equine Science and Management undergraduate intern, Univ. of KY, Lexington, KY (kaha258@g.email.uky.edu); Ext. Specialist, Univ. of KY, Lexington KY (raysmith1@uky.edu).

THE EFFECTS OF COVER CROPS IN AN INTEGRATED LIVESTOCK/CONTINUOUS CORN CROPPING SYSTEM IN EAST-CENTRAL MISSISSIPPI

K.R. Waddell, J.B. Rushing, B.S. Bass, and J.C. Lyles

There has been a growing interest on how cover crop systems affect soil health and economic productivity in integrated crop-livestock systems. This study will investigate the combined impacts of cover crop, tillage, and grazing in a continuous corn (*Zea mays*) production system. The study will be conducted at the Coastal Plain Branch Experiment Station in Newton, MS, on a Prentiss fine sandy loam soil. Tillage treatments will be applied prior to cover crop establishment. These will include conventional (CT) and no-till (NT). Twelve cover crop treatments consisting of mono- and polycultures of oats (*Avena sativa*), cereal rye (*Secale cereale*), crimson clover (*Trifolium incarnatum*), and radish (*Raphanus sativus*), along with a control, will be established in the fall of 2019 in two separate field studies: 1) grazed and 2) ungrazed. Plots are 25' x 30' in size. In the grazed plots, weaned beef steers (*Bos taurus*) will be used to graze cover crop treatments in a cafeteria style design fall through spring 2020. Forage mass, nutritive value, botanical separations, and selectivity will be measured. Corn will be planted following cover crop termination and stand counts and grain yield will be collected in summer 2020. Soil data collection will consist of soil moisture, compaction, bulk density, C/N, and nutrient availability. Comparisons will be made between grazed and ungrazed plots. Upon conclusion of the trial, a complete economic analysis including costs and net returns for each cover crop system will be calculated.

[1]Graduate Assistant, MSU-CPBES, Newton, MS (krw324@msstate.edu); [2]Assistant Research/Extension Professor, MSU-CPBES, Newton, MS (brett.rushing@msstate.edu); [3]Graduate Assistant, MSU-CPBES, Newton, MS (bsb341@msstate.edu); [4]Research Associate I, MSU-CPBES, Newton, MS (jcl411@msstate.edu).

THE EFFECTS OF GRAZING COVER CROPS ON ANIMAL PERFORMANCE, SOIL CHARACTERISTICS, AND SUBSEQUENT CROP PRODUCTION IN EAST CENTRAL MISSISSIPPI

B.S. Bass, J.B. Rushing, K.R. Waddell, and J.C. Lyles[1]

Integrated crop-livestock systems have been widely used throughout history, allowing for multiple agricultural practices on a single tract of land. With a growing awareness in recent years regarding soil health and agricultural sustainability, these systems have begun to be re-introduced into the southeastern U.S. Research will be conducted over a 2-year period at the Coastal Plain Branch Experiment Station (CPBES) in Newton, MS evaluating the impact of cover crops on animal performance, soybean (*Glycine max*) yield, soil physical properties, and overall economic productivity. Three cover crop treatments, replicated three times, were planted in a randomized complete block design on nine, two-acre paddocks on October 9th and 10th, 2019. The cover crop treatments were oats (*Avena sativa*), oats + crimson clover (*Trifolium incarnatum*), and oats + crimson clover + daikon radish (*Raphanus sativus*). Cover crops will be continuously grazed with commercial angus (*Bos taurus*) steers (± 500 lb) at 1000 lb/a stocking rate beginning in the fall of 2019. Steers will be weighed every 28 d. Average daily gain and total gain will be calculated for each cover crop treatment. Forage nutritive value will be analyzed on samples taken bi-weekly, with botanical separation readings taken monthly. Soil data collection includes soil moisture, compaction, bulk density, carbon/nitrogen analysis, and nutrient availability. An economic value will be placed on each system regarding net returns for animal performance and soybean grain yield.

[1]Graduate Assistant, MSU-CPBES, Newton, MS (bsb341@msstate.edu); [2]Assistant Research/Extension Professor, MSU-CPBES, Newton, MS (brett.rushing@msstate.edu); [3]Graduate Assistant, MSU-CPBES, Newton, MS (krw324@msstate.edu); [4]Research Associate I, MSU-CPBES, Newton, MS (jcl411@msstate.edu).

THE OPPORTUNITIES (AND CHALLENGES) OF GRAZING SUMMER ANNUAL FORAGE MIXTURES

K.M. Mercier, C.D. Teutsch, S.R. Smith, E.S. Vanzant, K.H. Burdine and E.L. Ritchey[1]

Abstract

Summer annual pastures provide an opportunity to add economical gain to weaned, fall-born calves, with the additional benefit of being able to sell at a more favorable late-summer market. In addition, summer annuals may be beneficial as part of a pasture renovation plan. Summer annuals are primarily grown in monocultures, however, past research with perennial grassland systems has shown benefits of increasing botanical diversity, an ecological principle that often leads to increased productivity. The objective of this study was to determine the impact of summer annual species diversity on the average daily gain (ADG) of weaned fall-born calves. Stocker calves (start weights of 725, 805, and 655 lb in 2017, 2018, and 2019, respectively) grazed one of three sorghum-sudan -based forage mixtures (monoculture = MONO, 3-species mixture = SIMPLE, and 12-species mixture = COMPLEX) for approximately 40 days during the summer months. The experimental design was a RCBD with three replications. Although several forage quality parameters were affected by forage mixture, none of these results provide useful insight into the differences observed in ADG. In 2017 and 2019, calves grazing MONO and SIMPLE mixtures had higher ADG than calves grazing COMPLEX mixtures (2017: 1.74 lb/day vs. 1.46 lb/day, $P < 0.03$; 2019: 1.3 lb/day vs. 0.95 lb/day, $P < 0.03$). In 2018, no differences in ADG were detected among treatments ($P > 0.3$); however, calves only gained 0.02 lb/day. Animal performance indicates that the increased seed costs to establish botanically diverse summer annual pastures may not be economical. However, increased species diversity may have other positive impacts in grassland ecosystems that were not measured in this study.

Introduction: The transition zone between the temperate northern and humid sub-tropical southern U.S. provides a unique environment that allows for the growth of both cool- and warm-season species. This allows for the production of high-quality forage nearly year around.

Growing cattle often require a higher plane of nutrition than cool-season perennial pastures can provide during the summer months. Tracy et al. (2010) found that summer annual pastures produced 61% more forage and exhibited equal or greater nutritive value as compared to cool-season pastures during the summer months. Therefore, summer annual pastures may offer an opportunity for retaining weaned calves during the summer months when cool-season forages are less productive.

[1] Graduate Student, University of Kentucky Research and Education Center at Princeton (UKREC), Princeton, KY (kelly.mercier@uky.edu); Extension Associate Professor, UKREC, Princeton, KY (chris.teutsch@uky.edu); Extension Professor, University of Kentucky, Lexington, KY (raysmith1@uky.edu); Associate Professor, University of Kentucky, Lexington, KY (evanzant@uky.edu); Associate Extension Professor, University of Kentucky, Lexington, KY (kburdine@uky.edu); Extension Associate Professor, UKREC, Princeton, KY (edwin.ritchey@uky.edu)

Increasing botanical diversity in native systems has been linked to increases in primary productivity (Hector et al., 1999). Past research has shown that summer annual mixtures of two species can out yield monocultures (Osiru and Wiley, 1972). However, limited data exist regarding the impact of species diversity in summer annual mixtures on animal performance. The objective of this study was to determine the impact of increasing species diversity on the average daily gain (ADG) of weaned fall-born calves.

Materials and Methods: This study was conducted in the summers of 2017, 2018, and 2019 at the University of Kentucky's Research and Education Center located near Princeton, KY (37.100664, -87.857355) (IACUC protocol 2017-2711). The soil type was a Crider silt loam. A randomized complete block design with three replications was used, with field position as a blocking factor. Nine two-acre pastures were sprayed with 2 qts glyphosate/A (one application in 2019 and two applications in 2017 and 2018 with two weeks between applications) to kill the cool-season vegetation. Pastures were then fertilized with triple superphosphate and potash to achieve target P and K fertility for warm-season annual forage crops according to the University of Kentucky's Lime and Nutrient Recommendations (Ritchey and McGrath, 2018). Ammonium nitrate was applied at rates of 60 lb N/A or 30 lb N/A, with the lower rate applied to pastures with treatments containing legumes. In late spring, pastures were sown with a no-till drill to one of three forage treatments: 1) sorghum-sudangrass (*Sorghum bicolor var. bicolor* × *bicolor*) monoculture (MONO), 2) SIMPLE mixture of sorghum-sudangrass, pearl millet (*Pennisetum glaucum* (L.) R. Br.) and soybean (*Glycine max* (L.) Merr.), and 3) COMPLEX mixture consisting of SIMPLE + corn (*Zea mays* L.), crabgrass (*Digitaria ciliaris* (Retz.) Koeler and *Digitaria sanguinalis* (L.) Scop.), cowpea (*Vigna unguiculate* (L.) Walp.), Korean lespedeza (*Kummerowia stipulacea* (Maxim) Makino), sunn hemp (*Crotalaria juncea* L.), sunflower (*Helianthus annuus* L.), forage rape (*Brassica napus* L.), and daikon radish (*Raphanus sativus* L.). Table 1 lists varieties and seeding rates used.

Stocker calves (start weights of 725, 805, and 655 lb in 2017, 2018, and 2019, respectively) were weighed on two consecutive days to obtain an average weight at the start and end of the study. Calves were randomly assigned and grazed one of three warm-season annual forage treatments without supplementation for 41, 36, and 48 days in 2017, 2018, and 2019, respectively. Calves were allowed access to ad libitum water and UK Beef IRM Cow-calf mineral. Paddocks were strip grazed in 2017 and 2018. In 2019, calves rotationally grazed paddocks that were divided in half. The first half was bush hogged after calves rotated to the second half. After calves had grazed both halves, hay was fed ad libitum until enough forage accumulated on the first half to sustain grazing.

Biomass was determined by clipping four 2.7 ft^2 quadrats leaving approximately 1 in. of stubble. A subsample for dry matter and nutritive value determinations was collected and dried for 5-7 days in a forced air oven at 130 °F. The subsample was ground through 2 mm and 1 mm screens sequentially using Wiley (Thomas Wiley, Philadelphia, PA) and Cyclone (Udy Corporation, Fort Collin, CO) mills, respectively. Nutritive value was estimated using near infrared reflectance spectroscopy, and total digestible nutrients (TDN) were calculated as follows: TDN = 100.32 – 1.118 * acid detergent fiber (ADF).

Results and Discussion: In 2017, calves grazing the MONO and SIMPLE mixtures had higher average daily gain (ADG) than calves grazing the COMPLEX mixture (1.74 lb/day vs. 1.46 lb/day, $P < 0.03$). In 2018, no differences in ADG were detected among treatments ($P > 0.3$); however, calves only gained 0.02 lb/day, most likely due to the advanced maturity of forages upon grazing initiation. In 2019, calves grazing the MONO and SIMPLE mixtures again had higher ADG than calves grazing the COMPLEX mixture (1.3 lb/day vs. 0.95 lb/day, $P < 0.03$) (Fig. 1).

In order to make comparisons with ADG, forage nutritive characteristics are presented by year in Table 2. In 2017, ADF, neutral detergent fiber (NDF), and total digestible nutrients (TDN) were affected by mixture ($P < 0.05$). In 2018, lignin, dNDF30 and IVTDMD30 were also affected by mixture ($P < 0.05$), and in 2019, no forage quality parameters were affected by mixture ($P > 0.05$). Interestingly, none of these results provide useful insight into the differences observed in ADG. It was also expected to see differences in lignin with each mixture, as each treatment had differing quantities of brown midrib sorghum-sudangrass planted, however, that was generally not the case. This implies that other factors besides forage quality measurements impacted ADG in this study.

One potential methodological fault was in sampling the entire plant for forage nutritive characteristics. Although it is understood that cattle do not evenly graze the sward to the ground, entire plants were sampled for simplicity and consistency. A better approach may have been to mimic the grazing patterns of animals used in this study. Therefore, animal performance did not reflect nutritive characteristics. In addition, dry matter intake was not measured in this study.

According to the University of Kentucky's Beef Ration Formulator, in 2017, protein supplied by forages should have allowed for ADG near 2 lb/day, meaning that cattle diets were energy limited. In 2018, protein and TDN were far below suggested values for growing steers to gain a target of 1.5 lb/day, and fiber concentrations were also high (40% ADF, 64% NDF). This likely limited intake and decreased digestibility, resulting in the poor gains observed in 2018. In 2019, although the Ration Formulator indicated that protein and energy were relatively balanced, the lower concentrations of these nutrients likely limited growth.

The results of this study indicate that increasing botanical diversity does not improve cattle performance. In fact, the complex summer annual mixture used in this study resulted in decreased ADG. However, these complex mixtures may offer other environmental benefits, such as providing habitat for pollinators. Future work should consider the impact of summer annual diversity on the entire grassland ecosystem.

Literature Cited:

Hector, A. et al. 1999. Plant diversity and productivity experiments in European grasslands. Science 286: 1123–-1127.

Osiru, D.S.O. and R.W. Wiley. 1972. Studies on mixtures of dwarf sorghum and beans (*Phaseolus vulgaris*) with particular reference to plant population. The Journal of Agricultural Science 79: 531-540.

Ritchey, E. and J. McGrath, editors. 2018. 2018-2019 Lime and Nutrient Recommendations. University of Kentucky Cooperative Extension Service bulletin AGR-1.

Tracy, B.F., M. Maughan, N. Post, and D.B. Faulkner. 2010. Integrating annual and perennial warm-season grasses in a temperate grazing system. Crop Sci. 50: 2171-2177.

Table 1. Forage species, cultivars, and seeding rates.

Mixture	Variety	Seeding Rate (lb/acre)	Seed Cost
Monoculture			
Sorghum-sudangrass	AS6402	50[a]	$79/A
Simple			
Sorghum-sudangrass	*Same as above*	25	
Pearl Millet	Wonderleaf	5	
Soybean	Big Fellow (2017) Large Lad (2018-2019)	25	
	Total	55	$86/A
Complex			
Sorghum-sudangrass	*Same as above*	10	
Sudangrass	AS9301 (2017) AS9302 (2018-2019)	4	
Pearl millet	*Same as above*	4	
Crabgrass	50:50 blend of Red River and Quick-N-Big	1	
Corn	AgriGold 115 day[b]	10	
Soybean	*Same as above*	10	
Cowpeas	Iron Clay (2017) Red Ripper (2018-2019)	10	
Korean lespedeza	VNS (2017 and 2019) Legend (2018)	4	
Sunflower	Peredovic (2017, 2019) VNS (2018)	2	
Rape	Barsica (2017-2018) T-Raptor (2019)	1	
Radish	Nitro (2017) SF Select (2018) Badger (2019)	2	
Sunn hemp	VNS	2	
	Total	60	$101/A

[a] High seeding rate used to ensure dense sward

[b] This variety was donated. Producer would be advised to use a cheaper hybrid; $2/lb was used to calculate seed cost.

Table 2. Mixture effects on forage nutritive characteristics by year.

		MONO	SIMPLE	COMPLEX	Mean	Significance
2017	CP	9.5 a	9.7 a	9.4 a	9.5	ns
	ADF	39.3 a	39.0 b	38.4 c	38.9	***
	NDF	64.1 a	62.7 b	62.2 b	63.0	*
	TDN	56.4 c	56.7 b	57.4 a	56.8	***
	Lignin	3.5 a	3.5 a	3.6 a	3.5	ns
	dNDF30	28.9 a	29.0 a	29.6 a	29.1	ns
	IVTDMD30	64.5 a	66.0 a	67.6 a	66.0	ns
2018	CP	8.2 a	7.2 a	6.5 a	7.3	ns
	ADF	40.1 a	39.6 a	39.8 a	39.8	ns
	NDF	65.6 a	64.0 a	63.3 a	64.4	ns
	TDN	55.5 a	56.1 a	55.9 a	55.8	ns
	Lignin	3.6 c	4.5 a	4.0 b	4.0	**
	dNDF30	28.5 a	26.2 b	28.4 b	27.7	**
	IVTDMD30	62.3 b	62.0 b	64.8 a	63.0	*
2019	CP	8.9 a	9.5 a	8.6 a	9.0	ns
	ADF	41.2 a	40.2 a	40.5 a	40.8	ns
	NDF	65.8 a	62.6 a	63.0 a	63.8	ns
	TDN	53.9 a	55.3 a	55.0 a	54.7	ns
	Lignin	3.2 a	3.2 a	3.3 a	3.2	ns
	dNDF30	32.0 a	32.5 a	32.3 a	32.3	ns
	IVTDMD30	67.7 a	71.5 a	70.6 a	70.0	ns

[a]Abbreviations: MONO = monoculture, SIMPLE = simple mixture, COMPLEX = complex mixture, CP = crude protein, ADF = acid detergent fiber, NDF = neutral detergent fiber, TDN = total digestible nutrients, dNDF30 = 30 h digestible NDF, IVTDMD30 = 30 hour in vitro true dry matter digestibility

[b] ns = not significant, * = < 0.05, ** = < 0.01, *** = < 0.001

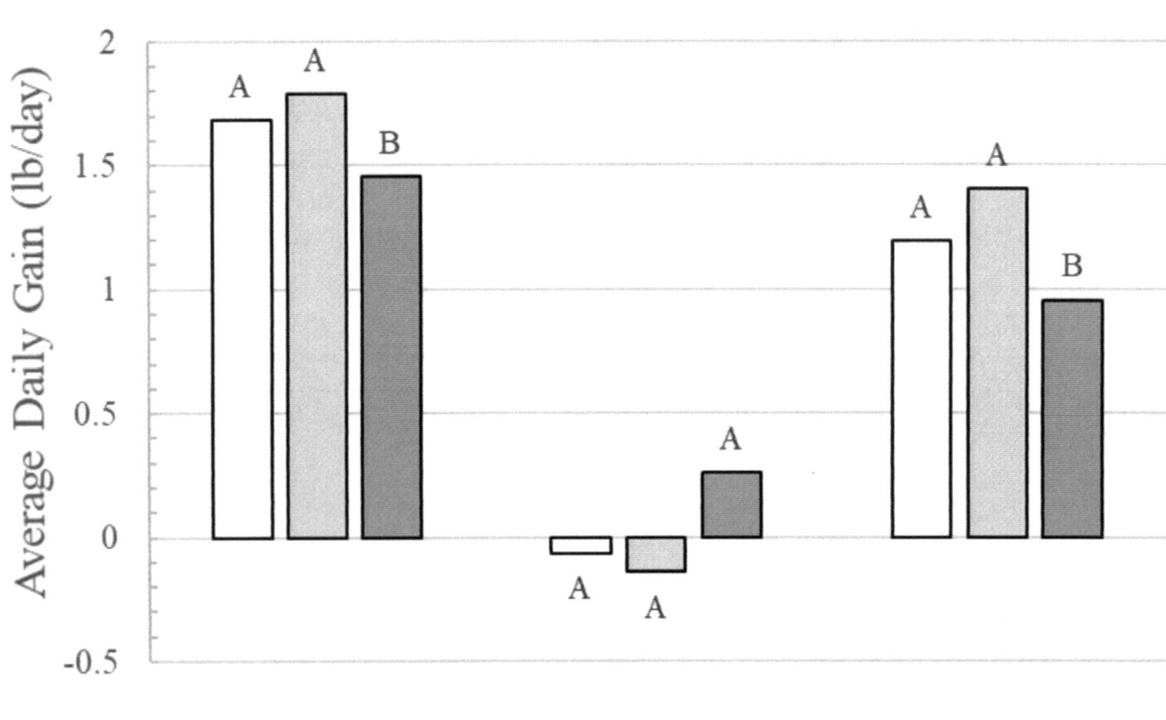

Figure 1. Average daily gain for stocker calves grazing a summer annual monoculture (white bar), simple mixture (light grey bar) and complex mixture (dark grey bar). Means within year with common letters do not differ ($\alpha = 0.1$).

THE SAFETY OF NATIVE WARM-SEASON GRASS HAYS FOR EQUINE

S. Ghajar[1], B. Tracy[2], K. Ely[3]

Abstract

Introduced cool-season grasses provide energy in excess of the dietary requirements of many horses, leading to obesity and an elevated risk of metabolic disorders such as laminitis. Native warm-season grasses are lower in energy and non-structural carbohydrates, making them potentially more nutritionally optimal for horses vulnerable to pasture-associated laminitis. However, some native forage species such as switchgrass (*Panicum virgatum*) and kleingrass (*Panicum coloratum*) are known to be toxic to horses, and little is known about the safety of other common native forage species. A toxicological trial was conducted to determine the safety of big bluestem (*Andropogon gerardii*) and indiangrass (*Sorghastrum nutans*) for horses. Nine Thoroughbred geldings were divided into three groups in a 3 x 3 crossover design and fed diets of exclusively big bluestem, indiangrass, and an orchardgrass (*Dactylis glomerata*) control for two weeks, followed by a two week washout period on pasture. Groups then rotated between treatments two more times with washouts in between each phase. Blood samples were collected prior to each phase, and weekly during feeding of the hay, and changes in albumin, aminotransferase (AST), bile acids, direct bilirubin, total bilirubin, sorbitol dehydrogenase (SDH), and triglycerides were compared. There were no differences between treatments. Big bluestem and indiangrass hay did not cause hepatic insult during the course of the study.

Introduction: Introduced cool-season grasses store excess photosynthetic products as non-structural carbohydrates (NSC), which are readily fermentable in the equine hindgut (Geor 2010). The high levels of NSC in introduced cool-season forages have been implicated in metabolic disorders in horses that precipitate laminitis, a painful and costly disease. Native warm-season grasses (NWSG), due their different photosynthetic pathway and growth form, have a different nutritional profile that may be more beneficial for equines. However, little research has been conducted to evaluate the use of NWSG as forages for horses. Some native grasses such as those in the Panicum genus are known to be toxic to horses, causing hepatic injury and potentially death (Lee et al., 2009). The potential toxicological effects of other common native warm-season forages are not known. This experiment examined the safety of two common NWSG species, big bluestem and indiangrass, for horses as hay.

Materials and Methods: The toxicological trial was conducted from November 2018 to January 2019 at Virginia Tech's Middleburg Agricultural Research and Extension Center. The experiment followed a 3 x 3 crossover design. Nine (n = 9) Thoroughbred geldings aged 6-15 were divided into three groups. Each group was fed exclusively one of three hays: big bluestem, indiangrass, or an orchardgrass control, in addition to ad libitum white salt and water. Hay was fed ad libitum for the first 8 days of each phase, then offered at 2.5% bodyweight daily to each horse during a concurrent digestibility trial. Blood samples were collected on Day 1 (baseline), Day 8, and Day 15 to evaluate biomarkers for toxicity, including albumin, gamma-glutamyl transferase (GGT), alkaline phosphatase (ALP), aspartate aminotransferase (AST), bile acids, direct bilirubin, total bilirubin, sorbitol dehydrogenase (SDH), and triglycerides. On Day 15, all groups were transferred to a cool-season mixed species pasture for a washout period of two to three weeks. Following the washout period, each group was exposed to another of the hay

[1]PhD Candidate, School of Plant and Environmental Sciences, Virginia Tech; [2]Professor, School of Plant and Environmental Sciences, Virginia Tech; [3]PhD Student, Department of Animal and Poultry Science, Virginia Tech.

treatments following the same protocol and washout period two more times to ensure each group had exposure to all three treatments.

Results and Discussion: Changes in biomarkers of toxicity did not differ on any of the biomarkers between treatment groups. One horse on the indiangrass treatment exhibited elevated biomarkers on the third blood sample during the first phase of the toxicological trial; however, the attending veterinarian attributed this to an unknown underlying subclinical issue rather than the hay, as no other horse exhibit a similar response at any point in the trial, and the values persisted after the horse was immediately switched to a washout diet. We conclude that big bluestem and indiangrass do not cause hepatic insult in horses after prolonged exposure. These forages may prove useful in feeding horses prone to obesity and laminitis.

Works Cited

Geor, R. J. (2010). Current concepts on the pathophysiology of pasture-associated laminitis. *The Veterinary Clinics of North America. Equine Practice*, *26*(2), 265–276. https://doi.org/10.1016/j.cveq.2010.06.001

Lee, S. T., Mitchell, R. B., Wang, Z., Heiss, C., Gardner, D. R., & Azadi, P. (2009). Isolation, characterization, and quantification of steroidal saponins in switchgrass (Panicum virgatum L.). *Journal of Agricultural and Food Chemistry*, *57*(6), 2599–2604.

THE SOUTHEAST HAY CONTEST: 2009-2019

Leanne Dillard[1], Marcelo Wallau[2], Lisa Baxter[3], Deidre Harmon[4], Dennis Hancock[3]

The Southeast (SE) Hay Contest was started by a group of Extension agents from across the SE with the objective of increasing producer knowledge of the relative forage quality (RFQ) index. Since its inception, the program has grown and in 2019 had over 350 entries from seven SE states. Forage samples were categorized into warm-season perennial grass hay (WSPGH), alfalfa hay (AH), perennial peanut hay (PPH), cool-season perennial grass hay (CSPGH), mixed annual grass, or other hay (MAG), grass baleage (GB), and legume baleage (LB). Samples were analyzed for nutritive quality using near infrared reflectance spectroscopy (NIRS) at the UGA Forage Testing Center (Athens, GA). The following summary includes data from 2009 – 2019. Total digestible nutrients were greater in LB and AH than all other forage categories (66.4 vs. 58.3%). The RFQ was greatest in AH (188) and least in WSPGH (106) over the 11-year period. Similarly, AH had the greatest CP (22.2%). However, WSPGH, MAGOH, and CSPGH had the least CP and were not different from each other (12.7%). As a result, RFQ was greater in AH than all other forage categories and least in WSPGH (188 and 106, respectively). While these forage samples were likely greater quality than samples not submitted, we conclude that outreach efforts, including the SE Hay Contest, have increased knowledge of the hay and baleage production process leading to increased hay quality from 2009 to 2018.

[1]Dept. of Animal Sciences/Dept. of Crop, Soil, and Environmental Sciences, Auburn University; [2]Agronomy Dept., University of Florida; [3]Dept. of Crop and Soil Sciences, University of Georgia; [4]Dept. of Animal Sciences, North Carolina State University.

THE UGA FORAGE EXTENSION TEAM IMPROVES FORAGE QUALITY AND PRODUCER LIFESTYLES TO ENHANCE SUSTAINABLE LIVESTOCK PRODUCTION

Morgan, S.[1], Knight, C.[2], Ray, L.[3], Kichler, J.[4], Gates, R.[5], Meeks, C.[6], Tanner, S.[7], Hancock, D.[8]

Sustainable forage-based livestock systems in Georgia have a farm gate value of over $1.4 billion. Sustainable agriculture is defined as an integrated system of plant and animal production practices having a site-specific application that will, over the long term: satisfy human food and fiber needs, enhance environmental quality and the natural resources base upon which the agricultural economy depends and enhance the quality of life for farmers and society as a whole. Forage quality is the key to a sustainable forage based livestock operation. As producers focus more on sustainability, Cooperative Extension must develop methods to meet the needs of these producers. The UGA Forage Extension Team has the goals to enhance sustainability the Forage Extension Team will positively influence and affect long term behavior change on the operations of Grassmasters graduates by reducing overstocking by 30%, assisting participants in attaining life goals and improving natural resources and ecosystems on the farm. In evaluating the long term success and behavior change of the program, 100% of respondents indicated that the program was helpful or a major help, 74% of respondents reported that they changed or adjusted their stocking rate, 68% reported that they changed their fertilizer application timing and 58% reported that they changed their forage testing and weed control program.

Harris County Extension Coordinator[1], Bulloch County Extension Agent[2], Morgan County Extension Coordinator[3], Colquitt County Extension Coordinator[4], Whitfield County Extension Agent[5], Houston County Extension Agent[6], Emanuel County Extension Agent[7], Extension Forage Specialist[8]

USING FORAGE BRASSICAS FOR RAISING STOCKERS AND IMPROVING SOIL CONDITIONS IN GEORGIA

Lauren Powell[1], Dennis Hancock[2], Tayler Denman[3], J.D. Hale[4], Edzard van Santen[5]

Forage brassicas have become a popular addition to winter grazing mixes, but there is limited research on the benefits. The objectives of this study were to compare a winter grazing mixture with and without forage brassica on forage productivity, animal performance, and soil health characteristics, such as changes in OM, compaction, and water holding capacity. A two-year grazing trial was conducted to compare performance of a mixture of rye, annual ryegrass, and crimson clover with and without brassica in a randomized complete block design with three replications. Experimental units were 5-acre paddocks with four tester steers and additional steers (grazers) used to adjust stocking rate. Forage samples were obtained every 28 d and were evaluated for quality. Likewise, steer weights were obtained every 28 d. Treatments showed no difference in forage yield early in the grazing season, higher yields for the mixture without brassica in late Feb./early March, and higher yields for the mixture with brassica in late Apr./early May. Steer average daily gains (1.23-3.76 lbs/head/d) were generally higher (by 0.53-0.76 lbs) for the mixture without brassica in the first half of the grazing season (Nov. – Feb.) but tended to be higher (by 0.48-0.87 lbs) for the mixture with brassica in the last two months (Apr. – May). Post-season soil bulk density was slightly higher after the grazing season ended in the paddocks containing brassica compared to pre-grazing samples, but likely of little practical significance and no differences were observed in areas where no brassica was used.

[1]Graduate Research Assistant – University of Georgia, Athens GA; [2]Professor and Forage Extension Specialist – University of Georgia, Athens GA; [3] Program Coordinator and Graduate Student – University of Georgia, Athens GA; [4] Research Technician – University of Georgia, Watkinsville, GA; [5]Agronomy Dept. and IFAS Statistical Consulting Unit – University of Florida, Gainesville, FL.

USING NATIVE WARM-SEASON GRASSES AND WILDFLOWERS IN PASTURES TO CONSERVE BEES

J. Wagner, M. O'Rourke, and B.F. Tracy[1]

In recent decades, habitat loss has caused many once-common native and managed bee species important to agricultural pollination to decline. Simultaneously, cattle producers in the eastern U.S. continue to experience issues with tall fescue toxicosis due to endophyte-infected Kentucky-31 tall fescue. In the fescue belt of the United States, seeding pastures with a mix of native warm-season grasses and native wildflowers could reduce tall fescue toxicosis issues while also increasing available resources for bees. In this study, a mix of three native warm-season grasses and 15 wildflower species was planted at the Virginia Tech Shenandoah Valley Agricultural Research and Extension Center. To measure the attractiveness of blooming flowers, all bee landings on flowering weeds and planted wildflower species were observed. We compared attractiveness of blooming weeds and flowers to bees between rotationally and continuously grazed cool-season grass pastures and pastures planted in native warm-season grass and wildflowers. The number of honeybees and native bees attracted to a plot increased with increasing bloom cover. Thistle was the most attractive weedy species, attracting an average of 2.23 ± 0.30 native bees and 0.13 ± 0.06 honeybees. In the wildflower pastures, planted flower species blooming mid-season attracted the highest average number of native bees (3.12 ± 0.47) and honeybees (3.00 ± 0.54). These results indicate that increased bloom cover, even from common pasture weeds, can increase available pollinator resources. Future research will examine the effects of native warm-season grasses and wildflower pastures on cattle production.

[1]Virginia Polytechnic Institute and State University, School of Plant and Environmental Sciences, Blacksburg, VA.

UTILIZATION OF GROWTH REGULATORS AND UREA ON DRY MATTER PRODUCTION OF COOL-SEASON FORAGES

Gelley, C.H.[1], Lima, D.F.[2], McCutcheon, J.S.[3], Turner, C.D.[4]

Gibberellins are compounds classified as plant growth regulators and has been shown to regulate multiple plant developmental factors such as breaking seed dormancy, and stem and leaf elongation. This study focused on increasing spring forage growth utilizing RyzUp SmartGrass®, a gibberellin based product, and fertilizer as dual growth stimulants during the spring of 2019. Previous data suggested a synergistic effect of RyzUp and urea; a follow-up experiment to test a low rate (0.4 oz.) and a high rate (1.0 oz.) of RyzUp was conducted. Four treatments were implemented; a control with no added urea or RyzUp, urea applied at 100 lb./ac., RyzUp applied at a 0.4 oz./ac., and both RyzUp and urea applied together in the early spring. A mirrored experiment with 1.0 oz of RyzUp was conducted concurrently. When tested at the 0.4 oz./ac. rate there were no significant differences among the treatments in biomass accumulated ($P = 0.10$). When tested at the 1.0 oz. /ac. rate of RyzUp in 2019, a significant increase in biomass accumulation was observed with both urea application and dual application of urea with RyzUp ($P < 0.05$). There was no significant difference between RyzUp applied at 1.0 oz./ac. and the control ($P > 0.05$). Under the conditions of this experiment, there was not a significant benefit observed by the application of RyzUp when compared to a traditional application of urea. Although stem elongation may have occurred, it did not translate to greater dry matter yield in this case.

[1] Agriculture & Natural Resources Educator, Ohio State Univ. Extension, Caldwell, OH, 43724 (gelley.2@osu.edu); [2] St. Clairsville, OH, 43950 (lima.19@osu.edu); [3] Assoc. Professor, Ohio State Univ., Caldwell, OH, 43724 (mccutcheon.30@osu.edu); [4] Agriculture & Natural Resources Educator, Ohio State Univ. Extension, Woodsfield, OH, 43793 (turner.1630@osu.edu).

UTILIZING COOL-SEASON ANNUAL GRASS AND GRASS-LEGUME MIXTURES FOR GRAZING AND SILAGE IN APPALACHIA

E. G. Henderson, D. D. Harmon, M. H. Poore, A. D. Shaeffer, and J. R. Rogers[1]

Opportunity exists for cool-season annual forages to extend the grazing season and provide high quality feed for beef cattle in Appalachia. The objective of this trial is to determine forage distribution, mass, nutritive value, and botanical composition of four cool-season annual grasses seeded with and without crimson clover. An additional objective is to determine whether inclusion of a homofermentative inoculant improves fermentation characteristics of cool-season annual grasses and grass-legume silages. Forage treatments of rye, ryegrass, wheat and triticale were seeded with and without clover in mid October of 2017. Half of each plot was harvested 4 times to mimic a grazing scenario while the other half of each plot was allowed to grow undisturbed and was harvested and ensiled with or without the inoculant. Averaged across all 4 harvests, grazing forage mass was greater for rye + crimson compared to triticale, wheat, and wheat + crimson (2,120, 1,685, 1,704, and 1,649 lbs DM/acre, respectively). Silage forage mass was greatest for triticale + crimson clover and least for ryegrass (10,306 and 6,703 lbs DM/acre, respectively), but did not differ among any other forage treatment. The results of our study indicate that producers should choose a winter-annual forage species that will best fill a forage gap in their production system.

[1]E. G. Henderson (eghender@ncsu.edu), D. D. Harmon (ddharmon@ncsu.edu), M. H. Poore (mhpoore@ncsu.edu), A. D. Shaeffer (adshaeff@ncsu.edu), J. R. Rogers (jrroger3@ncsu.edu), Dept. of Animal Science, North Carolina State University.

UTILIZING HORSE PASTURE FIELD DAYS TO EDUCATE HORSE OWNERS ON FORAGE MANAGEMENT AND PASTURE HEALTH

L. Ray[1] and P. Sapp

Horse owners comprise a unique segment of the livestock community. Horse owners can be individuals with minimal agricultural experience, which presents unique challenges when those individuals acquire land and set up a pasture system. Stocking rate, pasture rotation, and forage management are subjects that the equine industry can benefit from increased education. The Morgan County Extension Agent partnered with the Jefferson County Extension Agent to produce a Horse Pasture Field Day, held in Morgan County, in 2018 and 2019. The topics covered at these trainings include general horse grazing behavior, pasture layout and design, forage and soil testing, manure management, and hands-on pasture evaluation and weed identification. During the combined programs, 36 participants from eight counties attended the Horse Pasture Field Days. An evaluation to determine the effectiveness of these trainings found that 100% of participants agreed the information was presented in a method that was easy to understand: 1) 12.5% were satisfied with the relevance of the information received and 87.5% were very satisfied with the relevance of the information presented; 2) 87.5% of attendees indicated that they were very likely to implement methods learned at the training, 75% indicated that they would be more likely to conduct forage testing and 62.5% indicated that they were more likely to soil test as a result of this course. The results signify that these field days have positively affected the behavior of horse owners with minimal pasture acreages.

[1]Morgan County Agriculture and Natural Resources Agent, UGA Extension, Madison, GA (lray@uga.edu); Jefferson County Agriculture and Natural Resources Agent, UGA Extension, Louisville, GA (pamsapp@uga.edu).

WHERE DID THAT NITROGEN IN THIS PASTURE COME FROM?

A.J. Franzluebbers[1]

Fall stockpiling of tall fescue pastures is a key strategy towards more year-round grazing in the southeastern US. Standard recommendations are to stimulate fall growth of this cool-season grass with application of N fertilizer at the end of summer. But does the fertilizer actually yield forage and cattle benefits to cover the cost of fertilizer application? Might there be sufficient N already in soil to supply this crop with adequate N to make protein? How would we know if we didn't test the soil? A series of yield trials were conducted on farms throughout North Carolina (as well as in neighboring states of Georgia, South Carolina, and Virginia) with these questions in mind. Results suggest that soil N mineralization varies widely among farms and can be used as a basis for making N fertilizer decisions. Further, a simple and rapid test of soil biological activity from the flush of CO_2 following rewetting of dried soil effectively indicates the more complex assessment of soil N mineralization. Forage dry matter yield and quality will be presented and results across 92 field trials during past several years will help define an alternative N recommendation approach.

[1]Reseach Ecologist, USDA-Agricultural Research Service, NCSU Campus Box 7620, Raleigh NC 27695 (alan.franzluebbers@usda.gov).

WILDLIFE AND GRAZING

Mike Jones[1]

On our farm in northwest North Carolina, my wife Jean and strive to raise forage and cattle without being detrimental to the other species that live on the land. Wildlife is welcome always, no matter if prey or predator. Consisting of 215 acres, half of Beaver Creek Farm is dedicated to wildlife habitat and the rest to pastures of perennial and annual forages. Forages include warm-season and cool-season natives and introduced forages. Tall fescue dominates some paddocks and is not conducive to ground nesting song birds, such as Field and Grasshopper Sparrows. We offset the tall fescue with grasses and forbs that promote the existence of these small birds. Even in the fescue dominated paddocks, native grasses also grow. Large animals like White Tail Deer and Wild Turkey thrive on our place.

[1]Vice President (2019), North Carolina Forage and Grassland Council, forage and beef producer (bcangus@surry.net), Mount Airy, NC.

2010 Conservation Stewardship Award, Soil and Water Conservation Society, Hugh Hammond Bennett Chapter, and 2013 Conservation Stewardship Award, North Carolina Forage and Grassland Council.

YEAR THREE RESULTS OF NITROGEN OPTIONS FOR STOCKPILING COOL SEASON GRASS IN SOUTHEAST OHIO

J.S. McCutcheon, D.F. Lima, C.D. Penrose, M.A. Landefeld[1]

Abstract

Stockpiling cool season grass for grazing in the fall and winter is an option to extend the grazing season. The purpose of this study was to determine the effects of yield and quality by adding urea, urea with Agrotain®, and ammonium sulfate by applying them to cool season grass, primarily fescue and orchardgrass. This was the third year of the study and there were three sites in Southeast Ohio. Each site had a randomized complete block design with four treatments (control, 100 lbs./A urea, 100 lbs./A urea plus Agrotain®, and 219 lbs./A ammonium sulfate) and four replications of each treatment. The application date was August 6, 2018 and the plots were harvested on November 4, 2018. There were no statistical differences in CP, ADF, and TDN ($P < 0.05$). There were significant differences in yield between the control and all of the treatments, but not between the treatments. The three site average for the control was 2219 lbs. DM/A; urea, 2839 lbs. DM/A; urea+Agrotain®, 3190 lbs. DM/A; and ammonium sulfate, 2870 lbs. DM/A. Rainfall in the first 30 days from trial initiation in 2018 ranged from 3.21 to 3.73 inches, and the first significant rainfall (0.11-0.25 inches) was within 30 hours of initiation. This was the third and final year of this study and year three results continue to indicate that adding nitrogen increases yields.

Introduction: Many livestock owners use the granular form of urea nitrogen during late summer and fall trying to increase forage growth for "stockpiled" forage. Livestock are then allowed to graze the "stockpile" at a later date when other forages no longer are growing or available. This practice extends the grazing season and reduces the need for higher priced stored feed. This was an identical study to one that was conducted in 2016 and 2017 to replicate over time and was designed to determine the effects of using urea; adding a urease inhibitor (nitrogen stabilizer) product to urea, at the labeled rate, before applying the urea to the forage; and applying ammonium sulfate. The purpose of the study was to determine any difference in dry matter accumulation between treatments and detect changes in quality characteristics of the forages.

Materials and Methods: There were three locations in eastern Ohio (Monroe, Noble and Morgan Counties) with a randomized complete block design at each location with four (4) treatments, including a control, and four (4) replications of each treatment. Each plot was six feet by 20 feet. The fields were mechanically harvested prior to treatments to a height of three inches.

---

1 Associate Professor and Assistant Director, Ohio State University Extension; Extension Educator, Ohio State University Extension; Professor and Educator, Ohio State University Extension; retired Assistant Professor and Educator, Ohio State University Extension, respectively.

The control plots received no urea (46-0-0), urease inhibitor, or ammonium sulfate (21-0-0). For the other treatments, a total of 46 pounds/acre of nitrogen was used in each treatment in the following manner: 100 pounds urea/acre; 100 pounds urea/acre plus Agrotain® added at the labeled rate of one gallon* per ton of fertilizer; and 219 pounds/acre ammonium sulfate which was applied on July 31, 2017. The plots were harvested on November 13, 2017 to a height of three inches above ground level utilizing 2' x 2' subsamples from each plot. Each subsample was weighed fresh, and then taken to a laboratory for forage analysis. Each of the 48 samples was quality tested for Crude Protein (CP), Acid Detergent Fiber (ADF) and Total Digestible Nutrients (TDN). Statistics were calculated using Proc Mixed in SAS 9.3. Model included treatment, farm, and treatment by farm, with the random variable of rep within farm.

Results and Discussion: Table 1 lists the three site average for yield and quality. More detail for each specific county result and background information is available in the Fertilizer Options for Stockpiling Cool Season Grass (2018) technical report available at corn.osu.edu. Dry matter (DM) yields averaged 2645 pounds/acre for the plots where no nitrogen (N) was applied, 3322 pounds/acre for the plots with 46 pounds of N in the form of urea applied, 3494 pounds/ acre for the plots with 46 pounds of N in the form of urea applied, with Agrotain® added, and 3278 pounds for the plots with 46 pounds of N in the form of ammonium sulfate applied. There was a significant difference between the control and the treatments ($P<0.05$) for yield, but not between the treatments. There was also a significant difference in CP between the control and the treatments, but not between the treatments. There were no significant differences with ADF and TDN.

Research by McInnes, Ferguson, Kissel, and Kanemasu (1986) has demonstrated that urea nitrogen can be susceptible to volatilization under dry conditions when no rainfall occurs to move the broadcast nitrogen (N) into the soil in a timely manner. However, rainfall after the treatments were initiated occurred within 96 hours of the start of the study (0.7 in. Monroe Co., 0.75 in. Noble Co., 1.3 in. Morgan Co.) reducing the potential to lose N to volatilization. For the month of August, rainfall at the three sites ranged from 3.18 in. to 3.2 in.

There were no statistical differences in CP, ADF, and TDN ($P < 0.05$). There were significant differences in yield between the control and all of the treatments for the three site average, but not between the treatments. The three site average for the control was 2645 pounds DM/acre; urea, 3322 pounds DM/acre; urea+Agrotain®, 3494 lbs. DM/acre; and ammonium sulfate, 3278 pounds DM/acre.

Previous research conducted by Penrose (2015, 2018), and Landefeld (2016), showed a numerical increase in dry matter accumulation occurred using treatments over the control (urea alone and urea plus Agrotain®) even though there was no significant difference in the treatments at the ($P<0.05$) for yield. There was a significant difference in crude protein between urea (8.53%) and urea plus Agrotain® (8.31%) when compared to the control (6.77%) in the 2014 study. However, in the 2015 study, there was a significant difference between urea plus Agrotain® compared to the control and urea only. In the 2016 & 2017 portion of this trial, there was a significant difference between the control and the treatments for yield as in the 2018 portion of this trial.

One needs to calculate the application costs, consider the costs and time to feed stored feed, and the utilization of the stockpiled forages and the stored feed. In many cases, stockpiling is a viable option to reduce costs and save time.

Literature Cited:

Landefeld, M., Lima, D., McCutcheon, J., Penrose, C. (2019). Results of 2017 Fertilizer Options for Stockpiling Cool Season Grass. Available at: https://www.afgc.org/i4a/doclibrary/index.cfm?category_id=20&page=3

Landefeld, M. & Penrose, C. (2016). Comparing Fertilizer Additives Applied to Urea for Stockpiling Orchardgrass. Available at: http://www.nacaa.com/journal/post_editor.php?article_id=549

McInnes, K. J., R. B. Ferguson, D. E. Kissel, and E. T. Kanemasu. 1986. Field Measurements of Ammonia Loss from Surface Applications of Urea Solution to Bare Soil1. Agron. J. 78:192-196. doi:10.2134/agronj1986.00021962007800010038x

Penrose, C. Stockpiling of Forages. (2018). In 2018 eFields Report. Hawkins, E. editor. pp. 170-171. The Ohio State University. Available at: https://digitalag.osu.edu/efields/efields-reports/2018

Penrose, C., Landefeld, M., McCutcheon, J. & Lima, D. (2018). Fertilizer Options for Stockpiling Cool Season Grass. Available at: corn.osu.edu

Penrose, C., McCutcheon, J. & Landefeld, M. (2015). Utilizing Urease Inhibitors with Urea to Stockpile Fescue. Available at: http://www.nacaa.com/journal/index.php?jid=549

Table 1. Three Site Average

Treatments	lbs DM/acre	lbs DM/acre above control	CP%	ADF%	TDN%
Control	2219	-	9.4	30.0	67.0
Urea	2839	620*	9.8	32.0	65.3
Urea+Agrotain®	3190	971*	10.4	28.5	68.0
Ammonium Sulfate	2870	651*	10.1	31.7	65.8

LSD = 566 ($P < 0.05$) * denotes significant difference in yield compared to the control.

YIELD AND NUTRITIONAL COMPOSITION OF CONVENTIONAL AND BMR PEARL MILLET WITH DIFFERENT ESTABLISHMENT DATES OR HARVESTED AT DIFFERENT MATURITY STAGES

M. Oskey, B. Stancil and M. J. Aguerre[1]

Pearl millet (PM) is a drought-tolerant, warm season annual that adapts well to relatively infertile and acidic soils. Although there are several varieties with brown mid-rib (BMR-low lignin) trait, data on forage yield and quality is lacking. Two plot studies were conducted to determine the effect of planting date and maturity stage at cutting time on yield and quality of conventional and BMR PM varieties. In study 1, PM varieties were stablished on May 9 or 23 (10 plots per variety and seeding date) and harvested at late boot/early heading stage. In study 2, PM plots were established on May 10 and harvested at boot or heading stage (8 plots per variety and harvest stage). In each study, samples were collected from each plot and analyzed for nutrient content and in vitro true DM (IVTDMD) and NDF (IVNDFD) digestibility. In study 1, DM yield (lb/acre) from 3 cuts was 8.7% lower for BMR compared with conventional PM. Delaying establishment by 14 days reduced DM yield by 30% regardless of PM variety. Nutrient composition was similar between varieties (15.9% CP, 55.2% NDF, and 7.0% sugars). However, BMR had higher IVTDMD (82.0 vs. 80.9%) and IVNDFD (67.7 vs. 65.3%) compared to conventional. In study 2, DM yield from 2 cuts was 9.0% lower for BMR compared with conventional PM. Harvest at heading stage increased DM yield by 5.6%. Results of this study suggest that BMR PM is more digestible than conventional but at the expense of lower DM yield.

[1]Graduate student, Animal and Veterinary Science Department, Clemson University, Clemson, SC (moskey@g.clemson.edu); Program Coordinator, Clemson University Experiment Station, Clemson, SC (BSTANCI@clemson.edu); Assistant Professor, Animal and Veterinary Science Department, Clemson University, Clemson, SC (maguerr@clemson.edu).

YIELD, BOTANICAL COMPOSITION, FORAGE NUTRITIVE VALUE, AND PREFERENCE OF GRASS-LEGUME PASTURES UNDER HORSE GRAZING

Amanda M. Grev[1], Michelle L. DeBoer[2], Craig C. Sheaffer[3], M. Scott Wells[4], and Krishona L. Martinson[5]

Cool-season perennial grasses are the foundation of productive horse pastures in the North Central United States. The addition of legumes in pastures can increase forage yield and digestible energy content, yet little research has investigated grass-legume pastures under horse grazing. The objectives of this research were to evaluate yield, botanical composition, forage nutritive value, and preference of grass-legume pastures under horse grazing. Plots consisting of orchardgrass, Kentucky bluegrass, and meadow fescue were planted in monoculture and in binary mixtures with alfalfa or white clover and grazed by horses in Minnesota during the 2013, 2014, and 2015 grazing seasons. Yield, botanical composition, and forage nutritive value were determined prior to grazing, and horse preference was evaluated following grazing. When evaluating seasonal yield, meadow fescue and orchardgrass had greater yields than Kentucky bluegrass, and the inclusion of alfalfa resulted in consistently higher yields than grass monoculture or the inclusion of white clover. Additionally, the grass-legume mixtures resulted in a decrease in biomass of non-sown (weed) species, improved forage nutritive value, including higher crude protein and neutral detergent fiber digestibility, greater horse preference, and lower neutral detergent fiber. As a result, utilizing grass-legume mixtures in horse pastures can provide many benefits for grazing horses and pasture productivity.

[1]Extension Specialist, University of Maryland Extension, Western Maryland Research and Education Center; [2]Assistant Professor, University of Wisconsin - River Falls, Department of Animal and Food Science; [3]Professor, University of Minnesota, Department of Agronomy and Plant Genetics; [4]Assistant Professor, University of Minnesota, Department of Agronomy and Plant Genetics; [5]Professor, University of Minnesota, Department of Animal Science.

YIELD, NUTRITIVE QUALITY, AND SUGARCANE APHID INFESTATIONS OF SUMMER ANNUAL FORAGE MIXTURES IN CENTRAL ALABAMA

S. Thompson[1], A. Jacobson[2], J. Koebernick[3], K. Mullenix[1], G. Parks[1], and L. Dillard[1]

Summer annual forages [e.g., sorghum x sudangrass (*Sorghum bicolor* var. bicolor x *bicolor* var Sudanese; S)] provide a good option for producers to maintain beef cattle with greater nutritional needs with little to no external inputs due to their high yield and high nutritive value. However, S has recently had sugarcane aphid (*Melanaphis sacchari;* SCA*)* infestations, which damages the plant and reduces yields. In this study, mono-, bi-, and tri- cultures of -S, cowpea (*Vigna unguiculata*; C), and crabgrass (*Digitaria ciliaris*; R) were used and forage yield, quality, and SCA counts were determined for each treatment. Counts of SCA were conducted weekly and forage samples were harvested twice per season. Forage NDF and ADF was determined using wet chemistry at the AU Ruminant Nutrition Laboratory. During Year 1, the only difference in yield among treatments was observed in R, which was lower than other plots (949 and 1,259 kg/ha, respectively; $P < 0.05$). Neutral detergent fiber was greatest ($P < 0.01$) in treatments containing S, whereas C and C+R had the least NDF (64.6%, 55.1%, and 56.6%, respectively; $P < 0.02$). The SCA were greater ($P < 0.01$) in S+C compared to S+R and SI on July 12 (33.5, 21.1, and 22.4 SCA/ 2-leaf sample, respectively). During Year 2, there were no differences in yield among plots (996 kg/ ha, $P > 0.05$). Cowpea and C+R had the lowest NDF (46% and 45.6%, respectively, $P < 0.01$). While there appeared to be no reduction in SCA pressure in plots containing S, C+R showed promise as a summer annual mixture with similar yield and higher quality to that of S and S mixtures.

[1]Dept. of Animal Sciences; Auburn University; [2] Dept. of Entomology and Plant Pathology; Auburn University; [3] Dept. of Crop, Soils, and Environmental Science; Auburn University

2018—75TH YEAR SINCE RELEASE OF ICONIC COASTAL BERMUDAGRASS AND KENTUCKY 31 TALL FESCUE, 25TH YEAR FOR TIFTON 85 BERMUDAGRASS

G.M. Hill, V.A. Corriher-Olson and W. W. Hanna[1]

Kentucky 31 tall fescue (KY31) and Coastal bermudagrass (C) were released in 1943, and have been produced for over 75 years on millions of acres. KY31 continues to be used for grazing and hay production across the Upper South and lower Midwest, despite endophyte toxicity problems for cattle and horses. Consequently, research and management strategies have been publicized since the 1980s, resulting in release of fescue varieties containing non-toxic endophytes. Novel varieties alleviate toxicity problems in livestock while allowing higher performance in cattle and stand persistence in fescue fields. Both KY31 and novel fescues will continue to be major forages in these areas. Coastal Bermudagrass (C) is grown across the South extending from Virginia to Texas, and remains the standard for comparison of new bermudagrass selections. Tifton 85 bermudagrass (T85) was released in 1993. Compared with C, T85 has larger stems, wider leaves, and 15% to 25% greater yields with 10 % to 15% greater digestible dry matter. In a preweaning summer cow-calf study comparing C and T85 pastures, 91-d ADG was 19% greater for calves on T85 than C (2.07 vs 1.74 lb., $P < 0.01$), and T85 produced 5% higher weaning weights (557.6 vs 529.9 lb., $P < 0.05$). Both C and T85 are hybrids that must be established vegetatively, but T85 usually produces enough forage for light grazing or hay production in the establishment year. Adequate water, nitrogen fertilizer, and forage harvest maturity are key factors that affect quality of all bermudagrasses for grazing or hay production.

[1]Emeritus Professor, Animal and Dairy Sci. Dept., The University of Georgia, Tifton Campus, Tifton (gmhill@uga.edu); Associate Professor & Forage Extension Specialist, Texas A & M Agrilife Extension, Overton, TX (vacorriher@ag,tamu.edu); Professor, The University of Georgia, Formerly USDA-ARS Plant Geneticist, Tifton Campus, Tifton (whanna@uga.edu).

A PERSONAL EXPERIENCE: CREATING SILVOPASTURE WITH PRODUCTIVE FORAGES FROM AN OVERSTOCKED FOREST

J.D. Huff[1]

Silvopasture provides several production and environmental opportunities on farms that appeal to livestock producers such as expanding pasture acreage by thinning existing forestland. Cattle producers view this as an opportunity to adjust their current animal/forage balance to meet their production objectives. Based upon previous observations, creating a silvopasture system from an existing overstocked forest can be challenging. This presentation will share a personal experience of creating a silvopasture system with emphasis on applied techniques to reduce trees per acre, forage species selection and challenges that were presented. A 12-acre overstocked forest site in western Arkansas was chosen for converting into a silvopasture system. Specific hardwood trees were pre-determined to retain prior to thinning. A dozer with a shear blade was used to cut undesirables trees from the site in July 2018. The undesirable trees were piled and burned. Any remaining tree debris was removed from the site. The site was smoothed using a tractor and cedar tree. Annual ryegrass was broadcast planted in September 2018 for erosion control and forage. The annual ryegrass was mowed in June 2019. Pearl millet was no-till planted into the annual ryegrass residue and later mowed in October 2019 and planted again to annual ryegrass. Pictures and videos will be presented to show progression of the work.

[1]NRCS State Grazing Lands Specialist, USDA-NRCS (AR), jeremy.huff@usda.gov

A SIMULATED ECONOMIC ANALYSIS OF GRAZING WARM-SEASON ANNUAL FORAGES AS A CASH CROP ALTERNATIVE

Chris Prevatt, Kim Mullenix, Jennifer Tucker, Marcelo Wallau, James Yarborough, and Jane Cant

Row crop farmers in southeastern United States are searching for ways to further diversify their cash crop rotations and provide new income streams for their business as the current economic environment of low commodity prices offers few economic opportunities. One alternative that producers have shown interest in is the evaluation of integrating livestock into their cash crop rotation. A simulated economic grazing experiment was developed to evaluate the economics of grazing a warm-season annual forage crop over the last twenty years. Research data collected from the University of Florida, Auburn University, and the University of Georgia was used to evaluate the potential animal production, revenue, cost, and net returns for stocker cattle grazing warm-season annual forages. Production and economic variables evaluated for the warm-season annual forage system were stocking rate, number of grazing days, average daily gain, death loss, value of gain, cost of gain, and net value of gain. Over the time period evaluated, grazing a warm-season annual cash crop generated positive returns over total specified costs in seven out of the last twenty years. Therefore, producers would need to see additional benefits to future cash crops in order include grazing a warm-season annual forages in their cash crop rotation.

[1]State Specialized Agent, UF/IFAS Range Cattle Research and Education Center, Ona, FL (prevacg@ufl.edu); [2]Extension Specialist, Auburn University Department of Animal Science, Auburn University, AL (mullemk@auburn.edu); [3]Assistant Professor, University of Georgia Department of Animal and Dairy Science (jjtucker@uga.edu); [4]Extension Specialist, UF/IFAS Agronomy Department, Gainesville, FL (wallau@ufl.edu); [5]Agriculture Agent, UF/IFAS Orange County, Orlando, FL (jyarborough@ufl.edu); [6]Agriculture Agent, UF/IFAS Duval County, Jacksonville, FL (janeecant@ufl.edu)

AN INTERACTIVE EQUINE PASTURE MANAGEMENT WORKSHOP SERIES DESIGNED FOR THE FOOTHILLS REGION OF NORTH AND SOUTH CAROLINA

C.W. LeMaster[1]

The Carolina Foothills region is a unique area along the western border of North and South Carolina. State and county equine census data from 2004 and 2008 for the four neighboring counties (two on each side of the state line) range from 17,100 and 19,400 horses. The influx of non-native equestrians and farm managers to an already thriving equestrian community has led to greater demand for education covering best pasture management practices specific to our climate, soil and forage types. Clemson and North Carolina Cooperative Extension partnered to offer a unique interactive pasture management course for the Carolina Foothills area. The 1-d course, held in May 2019 at Cotton Patch Farm in Tryon, NC, offered 4 h of instruction. Participants were split and rotated between three groups to allow for hands-on opportunities. Of the 45 participants, 82% completed the written evaluation. Subject knowledge before and after the workshop was reported using a 5-point Likert scale (1 = very low, 5 = very high) in five categories: 1) determining body condition score, 2) effective and efficient grazing management, 3) importance of soil health, 4) sprayer calibration, and 5) identification of pasture plants and forage varieties suitable for local pastures. Reported subject knowledge rating increased in all 5 categories, with an average pre-workshop rating of 2.55 and an average post-workshop rating of 4.06. The survey responses indicate an eagerness among area horse owners to further their knowledge on the topics presented, and the collaborative Extension effort was an effective method to provide the best resources from each land grant institution.

[1]Area Livestock and Forage Agent, Clemson Cooperative Extension, Laurens, SC (cwycoff@clemson.edu).

ANIMAL PERFORMANCE IN GRAZING WINTER ANNUAL GRASSES WITH LATE MATURING CLOVERS

R. Lemus[1], J.D. Rivera[2] and J.A. White[3], J.G. Maples[4]

Winter annual forage systems (grasses and clovers) are critical in filling forage gaps in the southern USA and extending the grazing season. Since grasses and legumes mature at different rates, it is important to select species to balance forage biomass and nutritive values needed to maintain livestock production. Two grazing studies were conducted from 2015 to 2018 to determine the impact of late maturing clover and grass mix on livestock performance. Studies were conducted at the at the Henry H. Leveck Animal Research Farm at Mississippi State University. Experiment 1 (2014-2016) compared three treatments: annual ryegrass ('LoneStar') fertilized with commercial fertilizer (RN), annual ryegrass + 'Frosty' berseem clover (RGC), and 'Frosty' berseem clover alone (BC). Experiment 2 (2016-2018) compared three cool-season annual grasses (annual ryegrass, oat, and wheat) interseeded with 'Fixation' balansa clover. Grass monoculture was fertilized with 100 lb N/ac/yr. Treatments containing grass/clover mix were fertilized with 25 lb N/ac/yr after establishment. In Exp. 1, gain per acre (GPA) was greater for ryegrass with nitrogen (443 lb/ac) compared to ryegrass/clover (398 lb/ac) or berseem clover alone (116 lb/ac). No significant differences in GPA were observed in Exp. 2, although ryegrass clover had 2 and 6% higher GPA than oat and wheat, respectively.

[1]Professor, Extension Forage Specialist, Mississippi State University Extension Service (Rocky.Lemus@msstate.edu); [2]Associate Extension/Research Professor, South Mississippi Experiment Station, Mississippi State University (daniel.rivera@msstate.edu); [3]Forage Variety Testing Manager, Mississippi State University (jwhite@pss.msstate.edu); [4]Assistant Extension Professor, Mississippi State University (josh.maples@msstate.edu).

ASSESSING THE EFFICACY OF IN-FIELD BRIX MEASUREMENTS FOR FORAGE SUGAR AND ENERGY CONTENT

Eric Billman[1], Kathy Soder[1*], Jeff Horst[2], Kristi Balk[2], and Aimee Hafla[2]

Over the last decade, some dairy and livestock producers have been using the Brix index (a measure of sucrose concentration in solution) to assess forage energy content without having to rely on expensive and time-delayed wet-lab chemistry. A pilot study was conducted to test the accuracy of Brix results to actual sugar content in two common forage species. Four sampling periods occurred monthly from May through August 2019. At each sampling, eight alfalfa and orchardgrass samples were collected, with three Brix readings per sample measured on a digital Brix refractometer. Samples were flash-frozen, freeze-dried, and analyzed via wet-lab analyses for sugar content and nutritive value (Agri-King, Inc., Fulton, IL). Brix values were correlated with lab analyses in SAS and were either: a) negatively correlated to wet-lab analyses of total sugars (-0.65), glucose (-0.64), and fructose (-0.68) in orchardgrass, or b) not correlated at all to total sugars, glucose, and fructose in alfalfa. Conversely, Brix values of both orchardgrass and alfalfa were positively correlated to NDF (0.57 – 0.58) and hemicellulose (0.42 – 0.55), along with ADF of alfalfa (0.54). There was no correlation between Brix values and commercially-used metrics for forage quality, including TDN, RFV, and RFQ. These results indicate that the Brix index does not directly translate to forage quality. Producers should consider more accurate methods of analysis for assessing energy content of pastures.

[1]Research Animal Scientists, USDA-ARS, Pasture Systems and Watershed Management Research Unit, University Park, PA 16802 (Eric.Billman@usda.gov; Kathy.Soder@usda.gov)
[2]Laboratory Production Manager, Assistant Laboratory Production Manager, and Beef Cattle Nutritionist, respectively; Agri-King, Inc., Fulton, IL 61252 (Jeff.Horst@agriking.com; Kristi.Balk@agriking.com; Aimee.Hafla@agriking.com).

BEEF HEIFER DEVELOPMENT AND PERFORMANCE THROUGH BREEDING FROM ALTERNATIVE WARM-SEASON FORAGE SYSTEMS

M.L. Marks, C.G. Chappell, K.B. Mason, J. Yeager, R.B. Muntifering, S.L. Dillard, M.K. Mullenix[1]

Endophyte-infected tall fescue (ETF) is the predominant perennial forage used by beef cattle producers in the Black Belt prairie soil region of Alabama; therefore most beef heifers use this forage during the development phase of production. Summer perennial grasses could be an alternative forage to reduce the impacts of endophyte-infected tall fescue during heifer development. The objective of this study was to evaluate forage production and animal performance of weaned beef heifers grazing a mixture of big blue stem, little bluestem, and indiangrass (NWSG); Pensacola bahiagrass (PB) or ETF at the Black Belt Research and Extension Center, Marion Junction AL. Paddocks were continuously stocked with four weaned Angus x Simmental heifers from early June to late August 2018 and 2019. Put-and-take heifers maintained a target height of 15 in. for NWSG and 2-3 in. for PB and ETF. Forage mass was collected every two weeks and animal performance was measured every 28 d. There were no differences among forage systems for body weight, body condition score or average daily gain. Length of grazing between grazing systems differed: heifers grazed PB and ETF longer than NWSG. There were no yield differences between PB (4,795 lb./acre) and ETF (5,562 lb./acre), but NWSG (3,620 lb./acre) exhibited less production than the other species. Over the 2 years of this study, heifer gain was similar across various forage systems, demonstrating the potential Oof NWSG as an alternative to tall fescue in the Alabama Black Belt soil region. Additional research will follow heifers through breeding to evaluate potential conception and longevity differences.

[1]Auburn University/Alabama Cooperative Extension System, Auburn, AL

BIOMASS YIELD AND BOTANICAL COMPOSITION OF ANNUAL FORAGE MIXTURES FOR GRAZING

K. Mozea, M.T. Berti, K. Sedivec, A. Peterson, A. Wittenberg, S. Cabello, and A. Greenberg[1]

Growing forage annual mixture is a crop production technique that mitigates various aspects of farm practices, especially in the Upper Midwest region where the growing season duration is short. Forage mixtures could serve as a cover crop mix to protect the soil during the winter, for prevented planting areas, or as a high nutritious feed for grazing livestock. The objective of this study was to determine the botanical composition and productivity of selected annual forage mixtures. Twelve combinations encompassing 17 species were sown early summer (3 June 2019) at Fargo and Streeter, ND and the first harvest was on 1 August 2019. The mixture with the highest biomass yield (1.9 Mg/ha) included: foxtail millet (69%), forage sorghum blend (12%), oat (12%), forage pea (7%), and brassica hybrid-Winfred (1%). Percentages in parenthesis indicate the percentage of the crop in the mixture. The above mixture was different from monocultures of sorghum x sudan, brachytic sorghum (brown-mid-rib), foxtail millet, and forage pearl millet, with 1.4, 1.3, 0.8, and 0.8 Mg/ha, respectively. Using only 2.2 kg/ha of forage sorghum seed in a mixture did not reduce yield compared with mixtures with greater forage sorghum seeding rates. Grasses produced more biomass than dicotyledonous species in all botanical mixtures. These results emphasize how forage annual mixtures can provide additional fodder for livestock as well as act as soil cover in prevented planting situations.

[1]Graduate Research Assistant, MS student (kenneth.mozea@ndsu.edu); Forage and Biomass Crops Professor (Marisol.berti@ndsu.edu); Range Extension Professor (kevin.sedivec@ndsu.edu); Research Specialist (alan.peterson.2@ndsu.edu); PhD student (sergiofabian.cabello@ndsu.edu); MS student (alex.wittenberg.1@ndsu.edu); MS student (amy.greenberg@ndsu.edu) North Dakota State University, Fargo, ND

BOTTOM-LINE DECISIONS FOR FALL STOCKPILING

Greg Halich[1]

Discussions and debate on whether we should stockpile tall fescue pastures for winter grazing almost always center on nitrogen management: should we apply nitrogen and if so, how much. Studies have been extremely varied in terms of tall fescue forage response to nitrogen applications. Part of this variation can be explained by different climatic conditions of the studies (e.g. wet vs. dry years), but a main contention of this workshop is that a large part of this variation is likely due to previous studies not controlling for nitrogen that is naturally released through mineralization in the soil: Previous studies that showed large benefits to applying N are quite likely situations with low organic matter and biological activity in the soils, that led to little N mineralization. Previous studies that showed low benefits to applying N are quite likely situations where soil organic matter and biological activity were high. The other part of discussion on the overall decisions to stockpile tall fescue pastures that has also been largely missing is how to set up a farm to be able to stockpile. In most years, cool-season pastures have slowed down dramatically by August and forage availability is low. In these situations, it is difficult to pull pastures out of the grazing rotation to set aside and stockpile. This presentation will discuss strategies that allow pastures to be set aside, as well as the pros and cons of each: lowering stocking rates, using summer annuals or warm season perennials, and feeding hay (potentially supplementing pasture) during the stockpiling period.

[1]Associate Extension Professor, University of Kentucky, Department of Agricultural Economics, Lexington, Kentucky (Greg.Halich@uky.edu).

BREEDING FOR IMPROVED FIBER DIGESTIBILITY AND YIELD POTENTIAL IN TALL FESCUE (*Schedonorus arundinaceus* (Schreb.) Dumort).

E. Sahin[1], S. Reid[2], M. Sahin[1], S. Ates[3], S. Elias[1], D.B. Hannaway[1]

Tall fescue is a perennial forage grass widely used in areas not well-suited to perennial ryegrass due to climatic or edaphic stressors of drought and low or high pH. Although tall fescue has many agronomic attributes that make it well-suited to a wide range of environments, it is lower in palatability and digestibility than *Lolium* species. To take full advantage of tall fescue, genetic improvements in digestibility are needed to increase animal performance. The goal of this study was to improve fiber digestibility and biomass yield of medium-early maturing tall fescue accessions through recurrent phenotypic selection. Sixteen plants were selected from a source population of 1600 plants and placed in a polycross block. Using resulting half-sib seeds, a spaced-plant nursery was established in Philomath, Oregon to evaluate agromorphological characteristics. In addition, a small plot trial was established in Boyd, Kentucky in 2018 to quantify DM yield and estimate forage quality. Forage biomass, seed yield, plant height, tiller number per plant, and nutritive values were quantified. In the Oregon spaced-plant nursery, heading date was significantly different (P<0.05) among accessions, with Julian date ranging from 107 to 119 and seed yield ranging from 54.5 to and 21.3 g/plant. In the Kentucky small plot trial, DM yield ranged from 8,078 to 10,471 kg/ha. Future work will include additional selection cycles to further improve yield and digestibility.

[1] Graduate student, Department of Crop and Soil Sciences, OSU, Corvallis, OR (elif.sahin@oregonstate.edu); [2]Chief Breeder and Research Director USA DLF Pickseed USA, Corvallis, OR (sreid@dlfna.com); Graduate student, Department of Crop and Soil Sciences, OSU, Corvallis, OR (muhammet.sahin@oregonstate.edu); [3]Faculty, Department of Animal and Rangeland Science, OSU, Corvallis, OR (serkan.ates@oregonstate.edu); Faculty, Department of Crop and Soil Science, OSU, Corvallis,

OR (sabry.elias@oregonstate.edu); Faculty, Department of Crop and Soil Sciences, OSU, Corvallis, OR (david.hannaway@oregonstate.edu).

CHANGES IN BOTANICAL CANOPY COVER AND SEASONAL FORAGE PRODUCTION WITH HERBICIDE IMPREGNATED DRY FERTILIZER

Scott Flynn, Pat Burch, Michael Flessner, Wykle Greene, and Byron Sleugh[1]

Forage producers apply fertilizers and often herbicide treatments, but rarely both together. There is increasing interest in doing these two management practices at the same time to reduce costs. In addition, having increased fertility available as weeds are controlled will help desirable forage species respond and recover more once freed from competition with weeds. Experiments were set up over two years to assess the response of herbage mass and botanical cover following the application of dry fertilizer impregnated with a herbicide (DFI). Experiments were set up in 3 states and included multiple rates of DuraCor™ herbicide and GF-3731 at the appropriate timing for each location in a randomized complete block design with 4 replications. Data collected included monthly species composition from transect points to evaluate functional grouping such as forage grasses, forage legumes, weedy grasses, broadleaf weeds, sedges and rushes. Herbage mass of each vegetation class was determined. Each DFI treatment averaged greater forage grass herbage mass than the non-herbicide treated plots that received fertilizer only thus confirming that DFI is a good practice for achieving weed control and soil fertility to achieve a botanical shift to beneficial species in a sward.

Field Scientist, Corteva Agriscience, Lee Summit, MO (scott.flynn@corteva.com); Field Scientist (Retired), Corteva Agriscience, Christiansburg, VA (plburch@vt.edu); Michael Flessner, Assistant Professor and Extension Weed Science Specialist, Virginia Tech, Blacksburg, VA (flessner@vt.edu); Wykle Green, Grad student, Crop and Forage Weed Management, Virginia Tech, Blacksburg, VA (Wyke@vt.edu); Research Scientist, Corteva Agriscience, Indianapolis, IN (*byron.sleugh@corteva.com, corresponding author).

COLLABORATING TO FIND SOLUTIONS: A SUMMARY OF TWO YEARS OF ON-FARM BERMUDAGRASS STEM MAGGOT RESEARCH

Lisa Baxter, Holly Anderson, Andy Carter, Michasia Dowdy, Philip Edwards, Brooke Hampton, Stephanie Hollifield, Jeremy Kichler, Carole Knight, Will Lovett, Charlotte Meeks, Steven Powell, Pamela Sapp, Peyton Sapp, Andrew Sawyer, and Dennis Hancock[1]

The recent invasion of the bermudagrass stem maggot (BSM; *Atherigona reversura* Villenueve) has severely damaged bermudagrass (*Cynodon dactylon*) pastures and hayfields. Strategically-timed pyrethroid applications can significantly reduce adult BSM populations, but overuse of a single mode of action can lead to resistance. Cooperating Extension Agents/Coordinators and producers from 11 South GA counties conducted on-farm research from 2017 to 2019. The first trial consisted of screening potential alternatives insecticides. The second aimed to determine the best method for monitoring BSM activity. Since beginning this collaborative effort in 2017, we have observed an increase in producer adoption throughout South Georgia. Future research and outreach efforts will focus on expanding producer knowledge of integrated pest management strategies. Funding for this work was provided by the Georgia Beef Commission.

[1]Assistant Professor and Extension Forage Specialist, Crop and Soil Science Department (baxterl@uga.edu); University of Georgia Tifton Campus, Tifton, GA; Ben Hill County Extension Agent, Fitzgerald, GA (hhickey@uga.edu); Lowndes County Extension Agent, Valdosta, GA (aec22711@uga.edu); Brooks County Extension Agent, Quitman, GA (michasia@uga.edu); Irwin County Extension Coordinator, Ocilla, GA, (pedwards@uga.edu); Wheeler County Extension Coordinator, Alamo, GA (brooke@uga.edu); Brooks County Extension Coordinator, Quitman, GA (smh@uga.edu); Colquitt County Extension Coordinator, Moultrie, GA (jkickler@uga.edu); Bulloch County Extension Agent, Statesboro, GA, (clh@uga.edu); Manager at Griffin's Warehouse, Baxley, GA; Houston County Extension Agent, Perry, GA (cmote1@uga.edu); Treutlen County Extension Agent, Soperton, GA (steven.powell@uga.edu); Jefferson County Extension Coordinator, Louisville, GA (pamsapp@uga.edu); Burke County Extension Coordinator, Waynesboro, GA, (psapp@uga.edu); Area Pecan Agent Statesboro, GA (agsawyer@uga.edu); Forage Extension Specialist, Crop and Soils Science Dept., University of Georgia, Athens, GA, (dhancock@uga.edu).

COOL-SEASON GRASS HAY YIELD, FORAGE QUALITY AND CHANGES IN SOIL FERTILITY FROM POULTRY LITTER, BIO-SOLIDS AND COMMERICIAL FERTILIZER AMENDMENTS

N.R. Harris[1], T. Basden, E. Rayburn, J. Moore, B. Loyd, H.R. Scott, D. Bailey, J. Murray

The West Virginia poultry industry located in the Chesapeake Bay watershed produced 90 million chickens and 3 million turkeys, generating over 100,000 tons of litter (2016). To reduce nutrient applications in the Bay watershed, a poultry litter transfer cost share program has been initiated. The WVU Extension Service is conducting a response trial to evaluate poultry litter, commercial fertilizer and bio-solids to improve forage quality and soil fertility. The trial consists of utilizing high and low rates of poultry litter, commercial fertilizers at different rates that mimic the poultry litter nutrient values, bio-solids and a control. Soil samples 0-2 inches were collected each spring and fall from each plot. First hay harvest occurred June 7th, 2018 with a cutting width of 34 inches or 2.83 feet and the harvest length was 20 feet. Fresh weight from all plots was determined, a subsample was selected, dried, ground and analyzed for NIR 1 forage analysis. On June 7th second or split application on nitrogen was applied to High N fertilizer plots at 70 lb/acre N rate. A second harvest occurred on October 1 using the same procedure as the first. The results of the first year trial show a significant increase in annual yields for the high poultry litter and the high N commercial fertilizer treatments. Nutrient balance results showed that K was negative in all treatments, except for the high rate of poultry litter, high N-P-K, and Low N, high P-K. Results also showed that N balance was negative for low rate of poultry litter and low N commercial treatments. The bio-solid treatment had the highest N balance, but low yield, indicating that K is limiting bio-solid yield potential.

[1]West Virginia University Extension Service, ANR County Agent, Buckhannon, WV (natasha.harris@mail.wvu.edu)

DAMAGE AND RECOVERY OF FORAGES FOLLOWING RECORD ARKANSAS RIVER FLOODING

J. A. Jennings, K. J. Simon, M. S. Gadberry, K. Lawson, and J. Davis[1]

Flooding at record levels on the Arkansas River in spring of 2019 impacted over 76,000 acres of pasture and hayland. Producers needed information on the potential for forage recovery and need for reseeding. Relatively little information is available regarding forage recovery following prolonged flood events. Three fields were selected in central Arkansas to assess damage of existing forage and recovery over the growing season. The fields were initially submerged for 24 days followed by partial flood cover one week later. Primary forages were common bermudagrass, hybrid bermudagrass, bahiagrass, ryegrass, and johnsongrass. Drone images and ground assessments were made periodically through the season to monitor greenup and species percentage. Annual ryegrass was killed by the prolonged flooding. Both bermudagrass types and johnsongrass showed initiation of greenup within four days after the flood receded. Hybrid bermudagrass vigor was reduced much more than for common. Johnsongrass survived the first flood event but appeared to die out following the second flood event. Bahiagrass showed severe damage and little was found later in summer in flooded areas. Annual perennial grassy weeds invaded the hybrid bermudagrass sites and became dominant by the end of the growing season. Common bermudagrass had much less weed invasion. The order of recovery after flooding was common bermudagrass > hybrid bermudagrass > johnsongrass > bahiagrass > annual ryegrass.

[1]Professor, Forage Specialist, Little Rock, AR (jjennings@uaex.edu); Professor, Beef Specialist, Little Rock, AR (sgadberry@uaex.edu); Program Associate – Animal Science/Forages, Little Rock, AR, (ksimon@uaex.edu); CEA, Conway, AR (klawson@uaex.edu); Application Technologist, LFRS, Batesville, AR (jdavis@uaex.edu), University of Arkansas Division of Agriculture.

DECISION TOOL: ESTIMATED COSTS OF PRODUCTION FOR ORGANIC PASTURE MIXTURES

N. Colton, J. Allison, S. R. Smith, K. Burdine, K. Lea, D. Butler, G. Bates, G. Pigetti, A. Rius and P. Krawczel[1]

Pasture systems in the Southeastern U.S. are most commonly dominated by cool-season forages (i.e. tall fescue, red and white clover, or orchardgrass). During periods of elevated temperatures and decreased precipitation, the historical norm of summers in the Southeast, cool-season forage plant growth slows considerably. This phenomenon is often referred to as the "summer slump" because of the resulting impact that plant growth (or the lack thereof) has on grazing animal nutrition. Many livestock and dairy producers simply supplement decreased grazing intake with hay and concentrates. Dairy cows in organic systems must obtain at least 30% of their dry matter intake from pasture during the grazing season. This decision tool has been created with Excel to assist producers with calculating and analyzing the costs associated with complex forage mixtures for organic dairy system. Four proposed mixtures are included. They consist of cool and warm season annuals and perennials. The tool does not take forms of revenue into consideration, but considers input and machinery costs relevant to annual yields. All calculations are done on a per acre basis as well as at the whole-farm level. Individual sheets are provided for each budget, and a summary sheet is used to easily compare total costs across the mixtures.

[1] Plant and Soil Sci. Undergraduate, Univ. of KY, Lexington, KY (Nat.Colton@uky.edu); Ag. Econ graduate student, Univ. of KY, Lexington, KY; Ext. Specialist, Univ. of KY , Lexington, KY (Raysmith1@uky.edu);Research Analyst, Univ. of KY, Lexington, KY (krista.lea1@uky.edu); Associate Professor, Univ. of TN, Knoxville, TN (dbutle14@utk.edu); Ext. Specialist, Univ. of TN, Knoxville, TN (gbates@utk.edu); Associate Professor, Univ. of TN, Knoxville, TN (pighetti@utk.edu); Associate Professor, Univ. of TN, Knoxville, TN (arius@utk.edu); Associate Professor, Univ. of TN, Knoxville, TN (pkrawcze@utk.edu)

DEMONSTRATING CONVERSION OF WILDTYPE TO NOVEL ENDOPHYTE FESCUE PASTURES FOR GREATER LIVESTOCK PERFORMANCE AND BETTER ENVIRONMENTAL OUTCOMES

John H. Fike[1], Gabriel Pent, Scott Baker, Travis Bunn, Taylor Clarke, Amy Fannon, Lindy Fimon, Jeannie Layton-Dudding, Haley Norton, Morgan Paulette and Carl Stafford

Tall fescue is the primary forage base for Virginia's forage-livestock systems, and is widely planted because it tolerates low fertility and less-than-ideal management. A fungal endophyte in fescue supports grass growth and persistence, but produces toxins that negatively impact animal performance, partly by increasing body temperature. Current NRCS EQIP projects that exclude livestock from riparian areas to improve water quality may negatively impact animal welfare by limiting access to surface waters and shade. Novel, non-toxic endophytes can reduce animal stress and reduce animal use of streams and surface waters, but adoption has been slow. An NRCS-supported program was initiated to demonstrate the value of converting toxic fescue to non-toxic novel endophyte varieties. Initial surveys of 17 producer fields found that almost two thirds of fields had infection rates of 85% or greater. Educational programming is key to convincing producers to switch to novel endophyte fescue. Training sessions conducted with the Alliance for Grassland Renewal have reached over 100 producers and agents, defining the problem and presenting approaches for pasture conversion. On-farm demonstrations have demonstrated establishment methods and the benefits conversion to animal performance. Educating technical service providers has been an important component of these activities. Many have had limited knowledge of toxicosis and this program has helped them understand the need for fescue toxicosis mitigation in conjunction with stream exclusion.

[1]Associate Professor, Crop and Soil Environmental Sciences, Virginia Tech (jfike@vt.edu).

DESIGNING A PRECONDITIONING SYSTEM FOR RECEIVING STOCKER CATTLE USING COOL AND WARM-SEASON FORAGES

Chris Prevatt[1], Kim Mullenix[2], Jennifer Tucker[3], Marcelo Wallau[4], Jane Cant[5]
James Yarborough[6]

For one reason or another, many stocker cattle producers in the southeastern United States have struggled preconditioning feeder calves. An Extension project was designed to help producers with this implementation by developing a year-round 45-day forage preconditioning system for receiving stocker calves using cool and warm-season forages. This project helps producers understand the animal, forage, and resources needed to precondition feeder calves. The project was developed around the following assumption: each week one gooseneck load (approximately 20 head) of 350-pound steers are received on 120 grazeable acres. This project evaluated the required fencing, water, shade, animal health inputs, forage, and supplementation to meet and exceed the systems requirements. A major challenge producers face is providing adequate quantity and quality of forage and supplementation needed for year-round preconditioning. This was accomplished by utilizing cool-season annual forages (annual ryegrass, cereal rye, oats, etc.), cool-season perennial forages (tall fescue), warm-season annual forages (sorghum-sudangrass, pearl millet, etc.) and warm-season perennial forages (bahiagrass, bermudagrass, etc.). The total cost of implementing this year-round preconditioning system was $89/head over a 10 year period.

[1]State Specialized Agent, UF/IFAS Range Cattle Research and Education Center, Ona, FL (prevacg@ufl.edu); [2]Extension Specialist, Auburn University Department of Animal Science, Auburn University, AL (mullemk@auburn.edu); [3]Assistant Professor, University of Georgia Department of Animal and Dairy Science (jjtucker@uga.edu); [4]Extension Specialist, UF/IFAS Agronomy Department, Gainesville, FL (wallau@ufl.edu); [5]Agriculture Agent, UF/IFAS Duval County, Jacksonville, FL (janeecant@ufl.edu) [6]Agriculture Agent, UF/IFAS Orange County, Orlando, FL (jyarborough@ufl.edu)

DEVELOPING A YIELD ESTIMATION TOOL FOR ALFALFA-BERMUDAGRASS MIXTURES IN THE SOUTHEAST

K.M. Mason[1], K. Groce[2], M.K. Mullenix[1], J.J. Tucker[2], C.G. Prevatt[3]

Integration of alfalfa (*Medicago sativa* L.) into bermudagrass (*Cynodon dactylon* (L.) Pers.) stands is becoming increasingly popular in the Southeast. Sward height is an easy measurement for producers to determine yield in grass-based, monoculture systems. However, because species composition is variable in alfalfa-bermudagrass mixtures, yield estimation can be challenging. Combining estimations of stand variability, such as ground cover, botanical composition, and relative species contribution, with height and yield approximations could improve prediction estimates for mixed stands. A 2-yr simulated grazing trial was conducted in Shorter, AL and Tifton, GA, where 'Bulldog 805' alfalfa was interseeded into 'Tifton 85' bermudagrass. Plots were harvested at assigned interval (2, 4, or 6-wk) and height (2, 4, or 6-in) from June – October, 2018 and 2019. Ground cover (GC) ratings, botanical composition (BC), and dry-weight rank (DWR) were evaluated before harvest. Ground cover by alfalfa and other species or bare ground decreased throughout the year, while bermudagrass ground cover increased. The same pattern was observed for botanical composition and dry weight rank. For alfalfa, GC and BC were positively correlated ($R^2 = 0.8116$), as were GC and DWR ($R^2 = 0.6604$) and DWR and BC ($R^2 = 0.6855$). For bermudagrass, GC and BC were positively correlated ($R^2 = 0.8723$), as were GC and DWR ($R^2 = 0.5749$) and DWR and BC ($R^2 = 0.65$). Combining persistence estimates with yield and height measurements may lead to a more accurate yield estimation tool for producers.

[1]Department of Animal Sciences, Auburn University, Auburn, AL (kmb0108@auburn.edu), (mullemk@auburn.edu): [2]Department of Animal and Dairy Science, University of Georgia, Tifton, GA (jjtucker@uga.edu), (kaylyn.groce@uga.edu); [3]University of Florida, Institute of Food and Agricultural Sciences, Ona, FL (prevacg@ufl.edu).

DEVLOPING ANNUAL FORAGE ROTATIONS FOR HAY PRODUCTION IN THE SOUTHEAST

A.R. Burlingham[1]

In a climate of low commodity crop prices and increased sustainability, annual forages harvested for hay offer opportunities to create profitability while improving field conditions. Whether in rotation with commodity crops or in forage only rotations, selecting the right species and cultivars is important for harvest timing, weed and disease control and meeting market demands. Real opportunities exist in the southeast for growing a high value cool-season annual hay such as oats, or triticale between crops such as corn and cotton. Adding annual legumes to the mix can improve feed quality while creating opportunities to reduce fertilizer costs in the subsequent crop. Warm-season annuals offer alternatives to corn rotations on sandier soils between high value crops such as tobacco, sweet potatoes and peanuts. A multi-season rotation of forage crops between these high value crops can decrease soil born disease and pests, offer palmer amaranth control opportunities and improve soil health. While utilizing annual forages for hay production can be very beneficial, it is important to consider some of the challenges also present. Herbicide residuals from previous crops may be problematic and can lead to stand failure. Harvest timing and drying of annuals can also offer challenges to baling a quality feed. With understanding of the growth characteristics and harvest timing, annual forages can be included in crop rotations to increase net productivity per acre while improving whole farm sustainability.

[1]North Carolina Cooperative Extension Agent for Livestock and Forages, Pitt County Office, Greenville, NC (andy_burlingham@ncsu.edu).

DON'T FORGET ABOUT ORCHARDGRASS!

M. Poore, J. Rupert[1]

In recent years, there has been significant emphasis and energy surrounding the use of novel endophyte tall fescue. While important, the focus on a single species is at odds with recent interest in the planting of mixtures of species, and ideas about the benefits of species diversity in pasture. The singular focus on tall fescue has the potential to lead farmers to not plant other species that may be an equally good, if not better, fit for their environment; an example would be forgetting to consider a tried and true cool season forage – orchardgrass. Orchardgrass is one of the most versatile and broadly adapted of the cool-season grasses, is non-toxic and safe for all livestock, can withstand a wide range of soils, and is adapted to many regions of North America. Orchardgrass can be harvested by grazing, cut for storage as silage or hay, has excellent regrowth with multiple harvests, and it is well adapted to use in mixtures with other grasses and legumes. It is perennial providing many years of affordable forage with living plant material year-round, sustaining both animals and soil, and has been successfully used to partially renovate tall fescue pastures with pugging damage. Due to its shade tolerance, orchardgrass is a logical option for use in silvopasture systems, and is an agriculturally sound ground cover useful for buffer strips and reduction of soil erosion. While orchardgrass has limitations, particularly with overgrazing, cutting hay too short, susceptibility to selected pest and diseases, and limited heat adaptation in the Southern US, this species is one that should be part of a majority of our new planting and pasture renovation discussions. The benefits to producers are too many to ignore.

[1]Department Extension Leader and Ruminant Nutrition Specialist, North Carolina State University, Raleigh, NC (matt_poore@ncsu.edu); Marketing Development Manager, Smith Seed Services, Summerville, SC (jrupert@smithseed.com).

EFFECT OF ADDED POTASSIUM ON K-DEFICIENT TALL FESCUE-BROOMSEDGE HAYFIELD

Jimmy C. Henning, Joshua McGrath, Wade Turner, James Dollarhide and Gene Hahn[1]

Hay production removes large amounts of nutrients from the soil, especially potassium. Many producers commonly use 'balanced' fertilizers like triple-19 (19-19-19) with a 1:1 ratio of P_2O_5 to K_2O to fertilize hay. However cool season grass hays typically remove P and K in a 1:3 ratio. Therefore long term use of triple-19 for cool season grass hayfields can lead to K-deficient soils. In 2017, a producer consulted the University of Kentucky Cooperative Extension Service about a tall fescue hayfield which was underperforming. On closer inspection, the tall fescue stand had suffered severe encroachment by broomsedge (*Andropogon virginicus* L.). Soil tests revealed very low potash levels due to a history of fertilization of triple-19 at levels which did not equal K removal. Soil pH and phosphorus levels were adequate. A replicated small plot study was initiated in the spring of 2018 to study the effect of potash fertilization on forage yield. Plots were fertilized with treatments that included 38-38-38 (N-P2O5-K2O, lb/A) and all combinations of 0, 180; 0, 45; and 0, 180, 360 (N, P_2O_5, K_2O, lb/A). High fertility treatments (with N, P and K) increased total forage yield by 2.2 tons of dry matter (DM)/acre over the unfertilized control in the first season. Adding N+P+K changed the DM proportion of tall fescue and broomsedge from 11 and 89%, respectively for the unfertilized control to 86% and 14%, respectively. Treatments with P and K but no N stimulated the germination and emergence of volunteer red clover in the first and second year of the study.

[1]Extension Professor (jimmy.henning@uky.edu); Extension Professor (josh.mcgrath@uky.edu), Extension Field Program Coordinator (jwturner@uky.edu), Research Analyst (jedoll0@uky.edu), and Scientist I (gene.hahn@uky.edu), University of Kentucky, Lexington, KY 40546.

EFFECT OF DIFFERENT CUTTING SCHEMES ON YIELD AND FORAGE QUALITY OF LOW LIGNING AND CONVENTIONAL ALFALFA UNDER SOUTHEASTERN CONDITIONS

M. Oskey, B. Stancil and M. J. Aguerre[1]

High fiber concentration and low digestibility in forages limits dry matter intake because of increased feed bulkiness, slow clearance from the rumen, and reduced passage through the digestive tract. New reduced-lignin alfalfa varieties, recently available in the market, are a potential tool to increase utilization of forage with concomitant improvements in efficiency of nutrient utilization. However, information of yield and quality under southeast conditions are lacking. The objective of this study was to determine forage yield and nutritive value of reduced-lignin (RL) alfalfa vs. non-RL alfalfa cultivars with different cutting schemes. Alfalfa varieties (1 RL and 2 non-RL) were established in plots in the Fall of 2017 and harvested at 3 intervals (28, 35 and 42 days). Samples were collected from each plot and analyzed for nutrient content and in vitro NDF digestibility (IVNDFD). Yield (as-is) of RL was 2.7% and 20.2% lower compared with each of the non-RL varieties. Results from first cut samples indicated that RL had higher CP (24.4 vs 23.3%), but lower NDF (32.2 vs. 35.8) and lignin (4.9 vs. 5.9%) compared with non-RL. In vitro NDFD was 6.0% higher for RL compared to non-RL. Yield (as-is) of alfalfa cut every 28 days (6 cuts) was 2.7%, and 17% higher compared with 35 (5 cuts) and 42 (4 cuts) schedules. Preliminary results from this study suggested that RL has higher fiber digestibility and overall forage quality than non-RL, but at the expense of lower yield (as-is). These findings should be confirmed over several cuttings and seasons.

[1]Graduate student, Animal and Veterinary Science Department, Clemson University, Clemson, SC (moskey@g.clemson.edu); Program Coordinator, Clemson University Experiment Station, Clemson, SC (BSTANCI@clemson.edu); Assistant Professor, Animal and Veterinary Science Department, Clemson University, Clemson, SC (maguerr@clemson.edu).

EFFECTIVELY REACHING THE EQUINE COMMUNITY

Steven L. Musen, Krista L. Lea, S. Ray Smith and Robert J. Coleman[1]

Over the past 12 years, Extension Specialists and County Extension Agents from the University of Kentucky have put on an equine field day, The Kentucky Equine Farm and Facilities Expo. This annual event provides practical, hands-on demonstrations and seminars for more than 200 members of the equine community annually, particularly owners and managers of small and medium size horse farms, who may not have been previous consumers of Extension educational programming. One of the major draws of the Expo has been the host sites, which include various well-known Thoroughbred, Sport-Horse and Quarter Horse farms. The planning committee has also established partnerships with many equine businesses, recruiting over 30 companies to display their products and services at the event. The educational programs for the Expo have been conducted by Extension Specialist, industry experts and Extension Agents and have included topics such as; weed identification and control, vaccination protocols, farm safety, pasture management and renovation, feeding and nutrition, fencing, barn design, on-farm solar energy and many others. Post- field day evaluations have been conducted and, of the more than 2,400 persons who attended the field days, 84.25% of the respondents have found the information presented useful in the management of their horse operation, 78.5% planned to make at least one change in their horse operation and 90% were very satisfied with the event.

[1]Extension Agent for Agriculture and Natural Resources, Kentucky Cooperative Extension Service, Jessamine County, KY (steve.musen@uky.edu); Forage Research Analyst, University of Kentucky, Lexington, KY (krista.lea1@uky.edu); Extension Forage Specialist, University of Kentucky, Lexington, KY (raysmith1@uky.edu); Extension Equine Specialist, University of Kentucky, Lexington, KY (rcoleman@email.uky.edu).

EFFECTIVENESS OF INSECTICIDE APPLICATIONS ON BERMUDAGRASS STEM MAGGOT (BSM)

W.F Anderson Lisa Baxter, Dennis Hancock, and Will Hudson[1]

As part of the integrated pest management to mitigate bermudagrass stem maggot (BSM) (*Atherigona reversura* Villeneuve) more efficient and environmentally sound insecticide spray regimes need to be developed. The primary preventive spray recommendation is to spray two times with a pyrethroid and a Spinosad. Two forage bermudagrass cultivars (Tifton 85 and Alicia) were established in large (30' x 192') plots from sprigs to obtain a continuous stand. After full establishment, a 4-replication split plot RCBD design was employed with 8 spray treatments for each variety. Each treatment block consisted of a 6' wide strip for the full 30' length. The treatments were: 1) no spray; 2) spray once (10 days after clipping) with pyrethroid only; 3) spray twice (10 days after clipping and again 7 days after the first spray) with pyrethroid only; 4) spray once (10d) with Spinosad only; 5) spray twice (10d and 7d) with Spinosad only; 6) spray once with both pyrethroid and Spinosad; 7) spray twice with both pyrethroid and Spinosad; 8) spray both every week until harvest. Treatment 1 is the negative control and treatment 8 is the positive control. BSM appeared as early as June in 2017, 2018 and 2019. Dry matter yields significantly differed between treatments for clippings in July, August and September. Spinosad treatments were no better than the non-sprayed control and protection was not enhanced when Spinosad was combined with pyrethroids. Two sprays with pyrethroids was the best control but did not eliminate BSM damage.

[1]Research Geneticist, USDA-ARS, Tifton, GA (bill.anderson@usda.gov); Forage Extension Specialists, Crop and Soils Science Dept., University of Georgia, Athens, GA, (dhancock@uga.edu); (baxterl@uga.edu); Extension Specialist, Entomology Dept., University of Georgia, Athens, GA (wghudson@uga.edu)

EFFECTS OF NOVEL POTASSIUM SOURCES IN YIELD AND QUALITY OF ALFALFA

Y. C. Newman, W. Lenzen, A. Neitzke, and J. Bush[1]

High potassium (K) supplies are necessary in alfalfa for yield, quality and stand longevity. Alfalfa also has high requirements for the micronutrient boron (B), and in Wisconsin boron deficiencies are more widespread than that of any other micronutrient. Novel technologies combine K and B in one granule reducing the risk of uneven application given the small amounts required for B. Yield and nutritive value of novel potassium source (Aspire) compared to muriate of potash (MOP) were evaluated in Wisconsin on a two-year old stand of reduced lignin alfalfa (HarvXtra). Plots were cut four times per year during two seasons (2018 and 2019). There were five treatments which included one manured-only control (30 tons/acre), and four combinations of two sources of potassium (MOP or Aspire) applied at two rates [200 lb/acre/season after first cut (1X), or 400 lb/acre/season applied after first cut and again after third cut (2X)]. In 2019, seasonal yields for MOP at 1X or 2X were not different and averaged 7250 lb/acre (0.4 tons/acre increase over manured only). Similar trend was observed for Aspire, which averaged 7000 lb/acre (0.3 ton/acre increase over manured only) with no differences between low (1X) and high rate (2X). Under the conditions of this study, K application after first cut had a positive response (0.4 tons/acre increment over manured only with no K added) but no additional yield was observed with the second K application after third cut.

[1]Associate Professor/Forage Extension Specialist, Plant and Earth Science Department, Univ. of Wisconsin-River Falls, River Falls, WI; Research Assistant, Univ. of Wisconsin-River Falls, River Falls, WI; Research Assistant, Univ. of Wisconsin-River Falls, River Falls, WI; Research Assistant, Univ. of Wisconsin-River Falls, River Falls, WI

EFFECTS OF PLANT GROWTH-PROMOTING RHIZOBACTERIA ON STOCKPILED 'COASTAL' BERMUDAGRASS

M. L. Cole, M. E. Griffin, M. K. Mullenix, R. B. Muntifering, D. W. Held, S. L. Dillard[1]

Forage production practices have been greatly affected by the increasing cost of N fertilization. Supplemental and alternative N sources are needed to ensure economic viability of these systems. A 2-yr, small plot study evaluated plant growth-promoting rhizobacteria (PGPR) as an alternative form of N fertilizer for fall-stockpiled 'Coastal' bermudagrass. Eighteen 1-m^2 bermudagrass plots were treated with synthetic N fertilizer (N), DH44 (PGPR strain), DH44 + N, Blend 20 (PGPR blend), Blend 20 + N, a negative control, and stockpiled through the fall. Two applications of PGPR were applied in late-Aug and again 30 d later. Fertilizer and PGPR + N plots were treated with $(NH_4)_2SO_4$ at 56 kg N/ha. One-third of each plot was clipped to a height of 2.5 cm in Nov, Dec, and Jan. Forage DM yield, CP, NDF, ADF, and ADL were determined via wet chemistry. Forage DM yield was greatest ($P \leq 0.0318$) for Blend 20 + fertilizer, but was not different ($P = 0.2552$) from synthetic fertilizer (1,708 kg/ha and 1,577 kg/ha, respectfully). Concentration of CP was least ($P \leq 0.0437$) for DH44 and Blend 20 treatments (90 g/kg and 92 g/kg, respectively). Concentrations of NDF for control were different ($P \leq 0.0045$) from all treatments except synthetic fertilizer ($P = 0.1092$). Concentrations of ADF were not different ($P \geq 0.1613$), excluding the control ($P \leq 0.0525$; 342.8 g/kg and 358.0 g/kg, respectively). *In vitro* true digestibility was not different ($P = 0.0947$) among all treatments (463.1 g/kg). Yield and nutritive value parameters were greater ($P \leq 0.0246$) in yr 2 than yr 1. Results indicate that PGPR is a viable option for biofertilization of stockpiled bermudagrass; however, further investigation into the effects of PGPR inoculants on a larger scale is needed.

[1]Dept. of Animal Sciences, Auburn University, Auburn, AL (mlc0065@auburn.edu).

EMERGENCY FORAGE OPTIONS FOR MICHIGAN

K. Cassida, D. Baas, J. Islieb, M. Jean, P. Kaatz, M. Singh, K. Thurlow, and B. Wilke[1]

The last three growing seasons in Michigan and the Upper Midwest have presented unusual growing conditions, with a trend towards cold, wet winters with little snow cover, wet springs, and summer droughts. In 2019, weather extremes led to alfalfa winterkill, unprecedented scope of delayed planting for corn silage and other crops, and delayed first cut of hay and haylage, resulting in widespread shortage of good quality forage. Michigan producers are looking for alternative forage crops and management practices that can help buffer weather fluctuations. Winter annual small grains, such as triticale, rye, barley, or wheat, are seeded after silage corn harvest and can provide up to 2-4 ton/acre of forage in spring. Oats and peas planted in spring are another reliable option, yielding 1-3 tons/acre. Forage sorghum, sudangrass, and millets yield up to 6.25 tons/acre in Michigan, but Michigan does not reliably have enough growing degree days for maximum yield of these heat-loving crops. Early forage harvest of delayed planting cover crops was allowed for the 2019 emergency situation. Use of late-planted corn "cover crops" as emergency forage required adjustments to planting, harvest methods suitable for immature frosted silage, and monitoring for mycotoxins. Adaptation to changing weather conditions will require producers to innovate with adoption of these practices and more.

[1]Forage Specialist, Michigan State University, East Lansing, MI (cassida@msu.edu); MSU Extension Educator, Centreville, MI (baasdean@msu.edu); MSU Extension Educator, Munising, MI (isleibj@msu.edu); MSU Extension Educator, St. Johns, MI (atkinmon@msu.edu); MSU Extension Educator, Lapeer, MI (kaatz@msu.edu); Cropping Systems Specialist, MSU, East Lansing, MI (msingh@msu.edu); MSU Extension Educator, Gladwin, MI (thurlowk@msu.edu); Kellogg Farm Manager, Michigan State University, Hickory Corners, MI (wilkebro@msu.edu).

ENHANCED EFFICIENCY NITROGEN FORMULATION EFFECT ON TALL FESCUE – RED CLOVER MIXED SPECIES PASTURE PRODUCTIVITY

K.M. Payne, S.R. Smith, and B.M. Goff[1]

Nitrogen (N) fertilizer use is generally limited on mixed cool-season forages due to its stimulatory effect on grasses, increasing competition that can lead to reduced legume growth. The controlled-release nature of several enhanced efficiency (EE) fertilizer N products holds the potential to improve legume persistence in mixed species pastures while providing supplemental N to the grass component. This study evaluated the effect of different EE N formulations [Environmentally Smart Nitrogen (ESN), Agrotain®-treated urea (ATU), and a 75% ESN: 25% urea blend] and untreated urea on yield, nutritive value, and legume persistence in a 'KY 31' tall fescue and 'Kenland' red clover mixture in Lexington, KY during 2015-2016. Nitrogen was applied at four rates (0, 100, 200, and 400 lb N/acre) in three equally-split applications. During the first year, total seasonal yield increased with increasing N rates, but there was no difference in forage yield due to N rate or N source in 2016. The average red clover content increased to 68% by the end of 2015 and averaged 79% of the plot sward throughout 2016. Although clover content averaged 50% or more throughout the study, red clover in the sward declined with increasing N rate. Environmentally Smart Nitrogen and the ESN+urea blend maintained more red clover in the plots than ATU and urea. Forage nutritive value was not affected by N rate or N source. These results suggest that no N fertilization, regardless of N source, is required when red clover comprises more than 50% of the sward due to lack of yield or quality benefit.

[1]Applied Forage Systems Specialist, Virginia Tech, Blackstone, VA (kathrynpayne@vt.edu); Extension Forage Specialist, University of Kentucky, Lexington, KY (raysmith1@uky.edu); Agricultural and Natural Resources Extension Agent, West Virginia University, Point Pleasant, West Virginia (BGoff3@mail.wvu.edu).

ENHANCING BEEF CATTLE PROFITABILITY IN THE SOUTHEAST BY INCORPORATING WINTER-ANNUAL FORAGE

J.E. Cant[1] and C.G. Prevatt[2]

Hay supplementation is both an integral and expensive component of traditional annual beef cattle production in North Central Florida. Winter supplementation of hay for production females can exceed $300/head. The incorporation of cool-season annual forage into the grazing system can decrease annual cattle costs by up to $170/head by decreasing the quantity of supplements needed. The purpose of this program was to a) Educate producers on aligning cattle and forage production cycles to decrease winter supplementation costs and b) Demonstrate to producers the decrease in annual cattle costs by utilizing cool-season annual forages. Formal presentations in the morning covered weed management, economic management, establishment, and cultivar selection, and field visits in the afternoon provided hands-on experience with establishment, variety comparison, and stand persistence. Seventeen (17) producers representing an estimated 800 head of cattle attended the workshop. Based on evaluations, 78.2% (n=13) indicated an increase in knowledge on cool-season annual forage establishment, management, and utilization and 82.4% (n=14) indicated incorporating cool-season annual forages into their production systems. Attendees also estimated an average annual savings of $120.13/head by utilizing cool-season annual forages to extend their grazing season. This equates to $96,104 savings/year for the combined cattle represented. By decreasing production costs, producers can maintain sustainable cattle production more efficiently in an unstable market.

[1]Agriculture Agent, UF/IFAS, Duval County, Jacksonville, FL. (janeecant@ufl.edu)
[2] State Beef Cattle and Forage Economist, UF/IFAS, Ona, FL. (prevacg@ufl.edu)

ENHANCING THE PRECISION OF FROST SEEDING USING GPS

Conner Raymond and Chris Teutsch[1]

Guidance utilizing the global positioning system has long been used for various operations in row crop agriculture. However, the high price of these systems has limited use in low input forage-livestock operations. Reduced prices and the availability of used guidance systems has the potential to increase the use of precision agriculture in pastoral settings. In the past, overseeding clover into pastures during the winter months was a guess at best, often resulting in areas that received no seed and areas that were double seeded. The objective of this study was to evaluate the impact of using a guidance system on the uniformity of seed dispersal. This study was conducted at the University of Kentucky's Research and Education Center, located near Princeton. The experimental design was a randomized complete block with four replications. Four pastures ranging from 6.7 to 10.4 acres were overseeded (simulated) using an UTV equipped a Raven II Cruiser guidance system (Raven Industries, Sioux Falls, SD). The guidance system was initiated and then covered with a bag, and then the four pastures overseed by sight only. The process was repeated a second time utilizing the guidance system. Data was exported from the guidance unit in the form shapefiles. These files were analyzed for skips and overlaps using ARC GIS software (ESRI, Redlands, CA).

[1]Ag Technician and Associate Extension Professor, Dept. of Plant and Soil Sciences, Univ. of Kentucky, Lexington (chris.teutsch@uky.edu).

ESTABLISHING ALFALFA-BERMUDAGRASS GRAZING RECOMMENDATIONS IN THE SOUTHEASTERN U.S.

K.R. Groce, J.J. Tucker[1], K.B. Mason, L. Marks, and M.K. Mullenix[2]

Recently there has been growing interest in interseeding high-quality legumes, like alfalfa, into existing bermudagrass stands as a step towards improving forage, animal, and ecosystem stability in the Southeast. Current work evaluating these mixtures has focused on management and production of stored forages (i.e. dry hay/baleage), and limited work has evaluated the use of alfalfa-bermudagrass mixtures under grazing conditions. It is known that both grazing intensity and frequency impact long-term forage production. The objective of this research is to evaluate the effect of harvest frequency and height on newly established alfalfa interseeded into an existing bermudagrass stand and begin to develop preliminary rotational grazing recommendations for management of these mixtures. In 2018 thirty-six alfalfa-bermudagrass plots were established at 4 locations across Alabama and Georgia, using a randomized complete block design with four replications at each location. Alfalfa varieties were selected for suitability to location and were either 'Bulldog 505' or 'Bulldog 805' interseeded into existing hybrid bermudagrass sods. Treatments included combinations of harvest frequency (2, 4, or 6 weeks) and harvest height (2, 4, 6, inches). The mechanically harvested plots were evaluated for their influence on sward yield, persistence, stand density, botanical composition, and nutritive value. Depending on year and variety, harvest frequency, height, and their interaction effected yield ($P < 0.05$). Plots harvested every 4 weeks at 4 inches optimized yield and alfalfa persistence across the two seasons. This study is supported by funding from the USDA-NIFA-Alfalfa Forage Research Program (2017).

1Graduate Research Assistant, Dept. of Animal and Dairy Science (kaylyn.groce@uga.edu); Assistant Professor, Dept. of Animal and Dairy Science (jjtucker@uga.edu), University of Georgia-Tifton. [2]Graduate Research Assistant, Dept. of Animal Science (kmb0108@auburn.edu); Regional Extension Agent (mlm0013@auburn.edu); Assistant Professor, Dept. of Animal Science (mullemk@auburn.edu), Auburn University.

EVALUATING NITROGEN MANAGEMENT STRATEGIES TO IMPROVE SOUTHERN BERMUDAGRASS GRAZING SYSTEMS

J. C. Burt, J. J. Tucker, L. Baxter[1]

Bermudagrass (*Cynodon dactylon*) is one of the most common perennial warm-season forage crops grown in the South. Incorporating alfalfa (*Medicago sativa*) into bermudagrass pastures could serve as an alternative to synthetic N, while improving the nutritive value of the forage base. A two-year grazing evaluation was conducted from May-Sept. 2018 (Y1) and May-Aug. 2019 (Y2) at the University of Georgia Tifton Campus in Tifton, GA, to compare production of alfalfa/bermudagrass mixed pastures (BGA) with bermudagrass monoculture pastures with (BGN) or without (BGZ) synthetic nitrogen. Replicated paddocks were evaluated pre- and post-grazing for herbage availability, botanical composition, forage species competitiveness, and nutritive value. Paddocks (2 acres) were rotationally grazed using put-and-take management with two tester per treatment (stocker steers; Y1 BW=431.9±50.4 lbs.; Y2 BW=503.8±66.1 lbs.). Steers were weighed at initiation, conclusion, and on a 28-30 day interval for calculation of ADG and gain/ha. Despite significant drought in Y2, year did not affect total gain/ha, however treatment did (*P*=0.04), such that BGA was highest (342.2±31.3 lbs/acre), and BGN and BGZ were not different (233.0±31.3 lbs/acre and 213.2±31.3 lbs/acre, respectively). These data suggest that rotationally grazing alfalfa/bermudagrass mixtures can result in a higher gain/acre than bermudagrass pastures that are supplemented with or without synthetic N in the Southeast US.

[1]Graduate Student, Animal and Dairy Science Department, (justin.burt@uga.edu), Assistant Professor, Animal and Dairy Science Department, (jjtucker@uga.edu); Assistant Professor and Extension Forage Specialist, Crop and Soil Science Department (baxterl@uga.edu); University of Georgia Tifton Campus, Tifton, GA.

EVALUATING UNMANNED AERIAL VEHICLE (UAV) AND THERMOGRAPHIC IMAGING TECHNOLOGIES TO IDENTIFY GRASSLAND BIRD NESTS IN RECLAIMED NATIVE GRASSLANDS

L.M. Hearon and J.I. Morrison[1]

Grassland bird populations are declining at greater rates than any other group of birds. A well-established method of monitoring rapidly disappearing populations is by locating and identifying active bird nests. Studies quantifying grassland birds tend to have low statistical power due to low sample sizes, high labor costs, and high levels of disturbance - associated with difficulty finding nests. However, advances in unmanned aerial vehicles (UAVs) and thermographic imaging technologies have potential to improve efficiency of locating nests, while causing minimal disruption. Early research has evaluated nest detectability using a UAV equipped with a thermal imaging system. The UAV was flown at three different altitudes to detect simulated nests at incremental depths in monoculture grass stand canopies. This study evaluated nest detection accuracy using two different methods. The first required creating a flight plan for map construction of the research area using third-party software to later identify nests by heat signatures within the map. The second method required analyzing recordings of real-time video feed to identify and pin-point nest locations. Both methodologies were tested in a blind evaluation method, using multiple test subjects and replications. Results from this study have suggested that mapping software does not optimize nest detectability and identification. Future research will implement video analysis of UAV flights for nest detection in reclaimed native grasslands. This will allow for quantification of use and fitness of reclaimed native grasslands as habitat for grassland bird populations.

[1]Master's student, Dept. of Plant and Soil Sciences, Mississippi State University, (lm1358@msstate.edu); Assistant Research Professor, Dept. of Plant and Soil Sciences, Mississippi State University, (jim46@pss.msstate.edu).

EVALUATING UNMANNED AERIAL VEHICLE (UAV) AND THERMOGRAPHIC IMAGING TECHNOLOGIES TO IDENTIFY GRASSLAND BIRD NESTS IN RECLAIMED NATIVE GRASSLANDS

L.M. Hearon and J.I. Morrison[1]

Abstract

Grassland bird populations are experiencing a decline at greater rates than any other group of birds. A well-established method of monitoring rapidly disappearing populations is by locating and identifying active bird nests. Studies quantifying grassland birds tend to have low statistical power due to low sample sizes, high labor costs, and high levels of disturbance - associated with difficulty finding nests. However, advances in unmanned aerial vehicles (UAVs) and thermographic imaging technologies have potential to improve efficiency of locating nests with minimal disruption. Early research has evaluated detectability of grassland bird nests using a UAV equipped with a thermal imaging system. The UAV was flown at three different altitudes (30, 40, and 50 feet above ground level (AGL)) to detect simulated nests at incremental depths in monoculture grass stand canopies. This study evaluated nest detection accuracy using two different methods. The first was by creating a flight plan to allow for map construction of the research area using third-party software to later identify nests by heat signatures within the map. The second method required analyzing recordings of real-time video feed to identify and pin-point nest locations. Both methodologies were tested in a blind evaluation method, using multiple test subjects and replications. Results from this study have suggested that mapping software does not optimize nest detectability and identification. Future research will implement video analysis of UAV flights for nest detection in reclaimed native grasslands. This will allow for quantification of use and fitness of reclaimed native grasslands as habitat for grassland bird populations.

Introduction: Due to rapidly declining habitat, grassland birds are one of the fastest and most consistently declining group of birds in North America (Drum, 2015). Smaller populations have in turn created an increased chance of extinction, and concern for long-term population declines of grassland birds has led to an interest in how to monitor and manage them (Pimm et al., 1988). The ability to monitor grassland bird populations could aid efforts to reclaim and restore native grassland habitats needed for these populations to thrive. Grassland bird nest detection has proven to be a useful method in monitoring grassland bird population numbers (Martin, 1993). This research project evaluated using UAVs equipped with thermal imaging technology for grassland bird nest detection. Thermal imagery was collected and evaluated in two forms: live video or matrixed map. Further, this research sought to 1) identify strengths and weaknesses of currently available thermal imagery capturing and processing technology, 2) define accuracy and precision of human interpretation of those outputs, and 3) assess the validity and overall feasibility of this methodology. This novel technological approach to detecting grassland bird nests may prove to be quicker and easier than traditional detection methods. (Scholten, 2019).

[1] Masters student, Dept. of Plant and Soil Sciences, Mississippi State University, (lm1358@msstate.edu); Assistant Research Professor, Dept. of Plant and Soil Sciences, Mississippi State University, (jim46@pss.msstate.edu).

Materials and Methods: Research was conducted at Mississippi State University RR Foil Plant Science Research Center (33.467952, -88.754920) and HH Leveck Animal Research Center (33.423582, -88.792412) near Starkville, MS from 24 September to 6 December 2019. Simulated nest searches were conducted in five monoculture grass stands of giant miscanthus (*Miscanthus* x *giganteas* J.M. Greef & Deuter ex Hodkinson & Renvoize [*sacchariflorus* x *sinensis*]), big bluestem (*Andropogon gerardii* (L.) Vitman), Eastern gamagrass (*Tripsacum dactyloides* (L.) L., and two different ecotypes (upland and lowland) of switchgrass (*Panicum virgatum* L.). Technology evaluated consisted of the DJI Matrice 200 V2 UAV (S2 DJI Co-LTD, Shenzhen, China) equipped with DJI Zenmuse XT2 R 640 13 mm (30Hz) thermal camera. The UAV was flown at three different altitudes (30, 40, and 50 ft above ground level (AGL)) to detect simulated nests at incremental depths in monoculture grass stand canopies. Simulated nests were placed at ground level and ascending 1.5 ft increments until reaching maximum canopy height. Maximum canopy height between grass species ranged from three feet to seven and a half feet. Ten simulated nests were randomly distributed within each monoculture stand and placed at ground level. Three flights – one at each altitude – were conducted before nests were moved to the next canopy height. At canopy heights above ground level, nests were attached to plant stems using rubber bands.

Simulated nests were designed to meet a target nest temperature of 86°F, the mean temperature of typical grassland bird nests. Simulated nests were constructed by using standard HotHands® hand warmers placed inside Glad® 4oz. mini round food storage containers and secured with snap-on lids. When left in open air, hand warmers rapidly reached temperatures in excess of 100°F, however, when air was completely restricted, hand warmers failed to meet or maintain target temperature. In order to maintain consistent temperature, multiple designs which allowed circulation of fresh air within each nest were tested for consistency. The most reliable design from these trials resulted from perforating each nest lid with four holes, each approximately 0.25 in diameter. Hand warmers were removed from packaging, shaken for approximately 15 seconds, placed inside containers, and lids were applied. Nests were allowed to rest for 15 minutes, during which temperatures were recorded in five-minute intervals. On average, simulated nests reached target temperature within nine minutes and maintained target temperature for over three hours. The maintenance of temperature was crucial in order to accurately portray grassland bird nest temperatures and accurately determine the usefulness of UAV/thermal technology for nest detection.

Flight plans for mapping each grass monoculture were created using DroneDeploy® software prior to arriving on scene. All flight plans were programmed with specific flight altitude (30, 40, or 50 ft AGL), front and side overlap (90%), and flight speed (5mph). Thermal videos were recorded simultaneously during flight with the application MyScreenRecorder®. The two imagery capture applications were launched and operated simultaneously throughout flight.

Both thermal maps and videos were analyzed by a blind review process. All identifiable information (monoculture species, flight elevation, canopy height, and date) was removed during evaluation of imagery. Including the pilot, a total of five test subjects were trained to identify heat signatures in thermal imagery. Individually, subjects were given all maps - arranged in a random order - and asked to analyze each map for heat signatures. This process - with identifiable information removed and in random order - was repeated for video analysis. Data were compiled and analyzed using PROC GLM in SAS® statistical analysis software, version 9.4. Test subjects, flight altitude, depth in canopy, and grass species were all treated as dependent variables. All main effects and interactions were considered significant at $\alpha = 0.05$ level.

Results and Discussion: Results indicated that accuracy and precision are significantly increased when analyzing live video feed as opposed to matrixed maps to pinpoint nests. Overwhelmingly, altitude had little impact on accuracy or precision of detection regardless of imagery type, while depth in canopy (DIC) had significant impact on accuracy of detection when using matrixed maps. Images 1 and 2 help illustrate the variability in quality of constructed maps. At a flight altitude of 50 ft (Image 1), images matrixed well and maps generally yielded high simulated nest detectability. At an altitude of 30 ft (Image 2), however, image stitching quality was generally sub-optimal, and the number of positive nest detections declined.

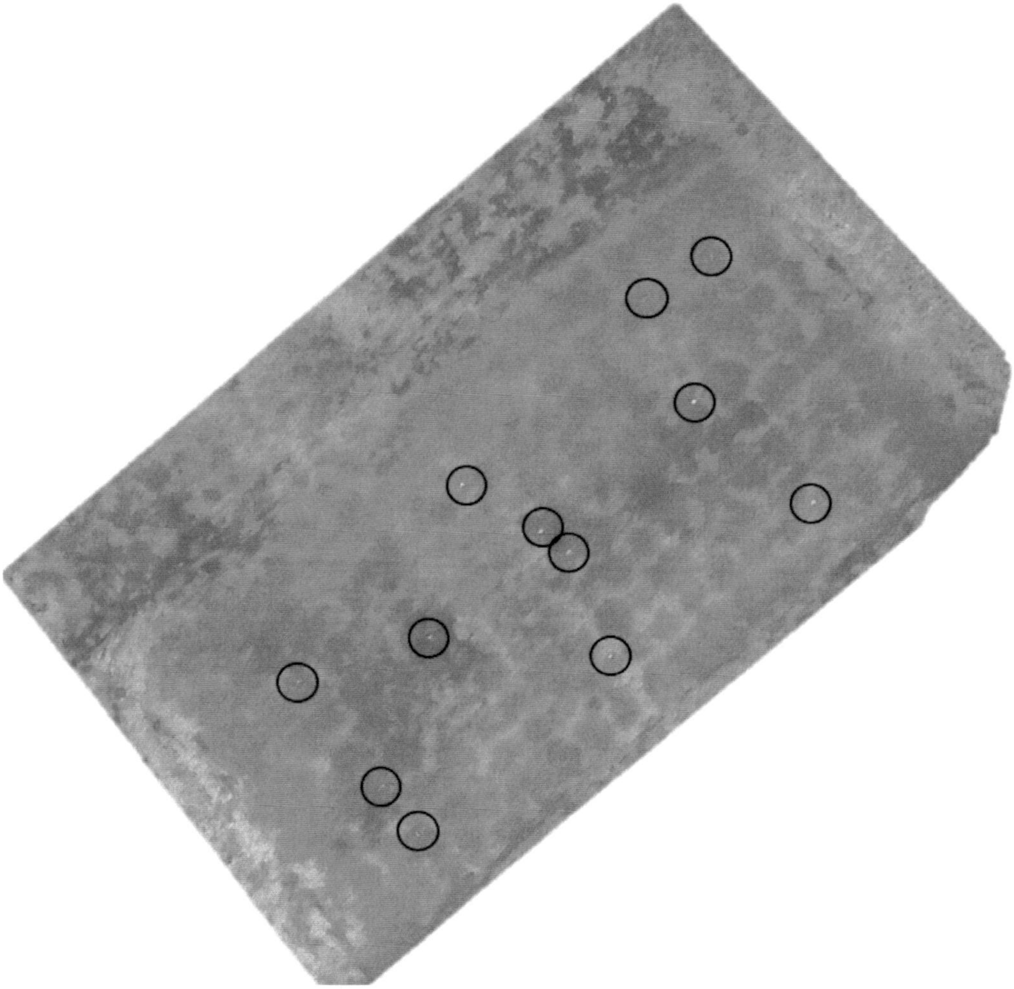

Image 1. Thermal map generated from imagery collected in upland switchgrass monoculture, at 50 ft AGL, with simulated nests at a depth of 1.5 ft in canopy. Outlined are 12 probable simulated nests out of ten that were placed in the plot.

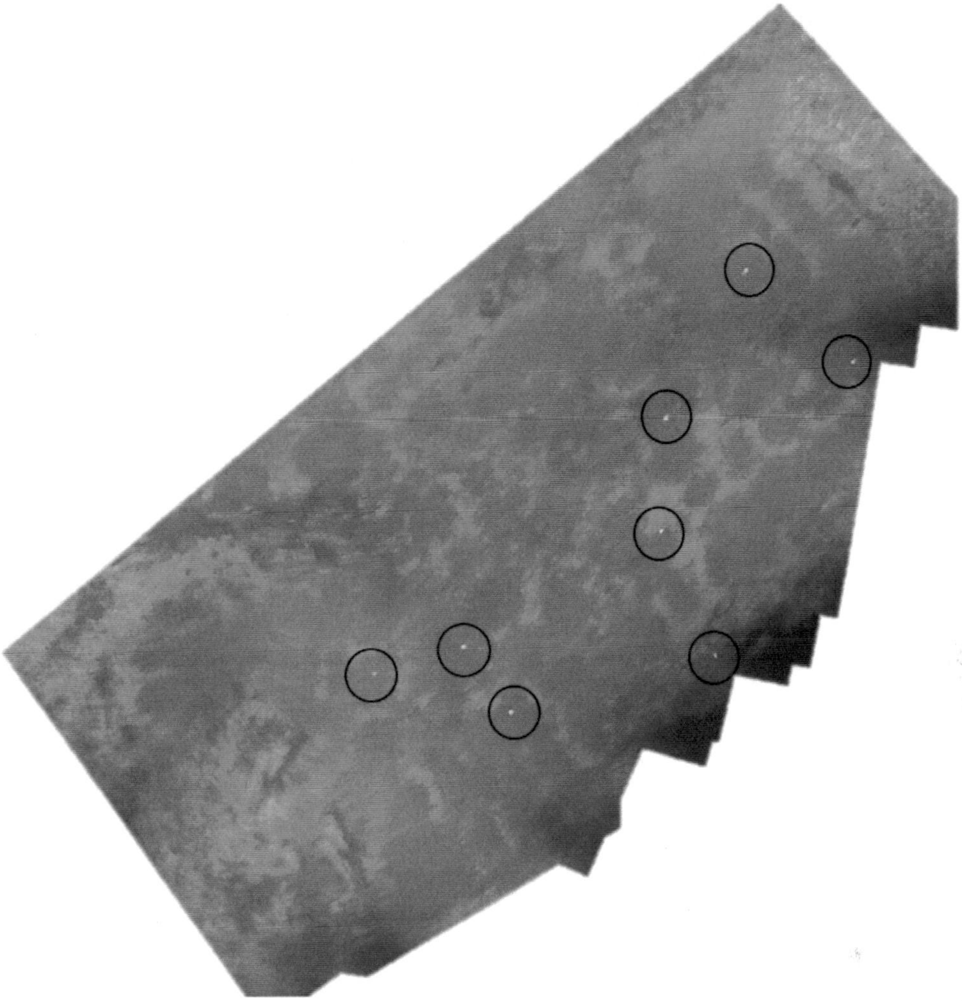

Image 2. Thermal map generated from imagery collected in upland switchgrass monoculture, at 30 ft AGL, with simulated nests at a depth of 0.5 ft in the canopy. Outlined are eight positively identified simulated nests out of ten that were placed in the plot.

While test subjects indicated that thermal maps are ideal for timely analysis of imagery, currently, mapping software is not capable of accurately translating thermal imagery into maps at low flight altitudes. It was assumed that low flight altitudes would be required for researchers to positively identify nests. However, current hypotheses propose that low flight altitudes lead to inadequate geospatial data, causing mapping algorithms to fail to create maps with correct overlap of images. Currently, maps from this research tend to have areas of duplication, blur, and distortion caused by the mapping algorithm used. These inconsistencies inflate inaccuracy and imprecision in nest estimation.

Literature Cited:

Drumm, R. G., C.A. Ribic, K. Koch, E. Lonsdorf, E. Grant, et al. 2015. Strategic grassland bird conservation throughout the annual cycle: linking policy alternatives, landowner decisions, and biological population outcomes. PLoS ONE 10(11): e0142525. Pmid:26569108.

Martin, T. and G. Geupel. 1992. Nest-monitoring plots-methods for locating nests and monitoring success. J. Field Ornithol. 1993.

Pimm S.L., Jones H.L. and Diamond J. 1988. On the risk of extinction. Am. Nat. 132: 757–785.

Scholten, C. N., A. J. Kamphuis, K. J. Vredevoogd, J. L. Atma, and C. B. Shea. 2019. Real-time thermal imagery from an unmanned aerial vehicle can locate ground nests of grassland songbird at rates similar to traditional methods. Biological Conservation 233:241-246.

EXTENSION PROGRAMS FOR NEW AND BEGINNING FORAGE PRODUCERS

M.D. Mauldin[1]

The rural inland areas of northwest Florida are experiencing a steady influx of individuals moving from southern/coastal areas of Florida or from colder parts of the country. Many of these individuals are embracing their new rural settings and attempting to begin livestock operations. Unfortunately, many lack the knowledge and experience necessary to realize their goals. This is particularly true in relation to forage production. In response, University of Florida Institute of Food and Agricultural Sciences (UF/IFAS) extension faculty planned and implemented a variety of programs designed to help beginning forage producers become more successful. The following illustrate some key efforts by the UF/IFAS Extension Agent in Washington County, both as an individual and in association with Panhandle Ag Extension Team: *Principals of Pasture Productivity* and *Beef Cattle Basics* – two series developed and taught by the agent specifically for new and beginning producers; 22 separate events, total attendance 360, 99% of respondents reported knowledge gains (n=226), and 75% of respondents reported planned practice changes (n=117). *Northwest FL Grazing School* – one day event, planned and presented with the Panhandle Ag Extension Team, total attendance 64, 95% of respondents reported knowledge gains (n=60), and 71% of respondents reported planned practice changes (n=45). From 2015 – October 2019, the agent conducted approximately 1,340 individual consultations specifically related to forage production, performed four on-farm pasture herbicide demonstrations, and 3 on-farm forage variety demonstrations (1,234 instances of knowledge gain and 966 practice changes).

[1]Agriculture & Natural Resources Agent, UF/IFAS Extension Washington County, Chipley, FL (mdm83@ufl.edu).

FACILITATING UNDERGRAD RESEARCH – CLOVERS FOR POLLINATORS

Roger Gates, Michael McClurken, Brittany Jensen, Jared Burns and David DesRochers[1]

Invertebrate pollinator numbers have declined due to a variety of factors: pesticide exposure, habitat fragmentation and loss, and disease. Forty-six species of invertebrate pollinators have been listed as endangered in the U.S. Large areas of municipal turfgrass receive limited or no fertilizer or pesticide, but are mowed regularly, providing a suitable environment for clover, which could provide floral resources to enhance pollinator habitat. Two projects were implemented to determine how well planted flowering resources do in minimally managed areas, estimate diversity of invertebrate pollinator communities present, and provide opportunity for undergraduate students to conduct field research with vegetation. In 2017-18, 3 reps (10' X 10') of 7 clover entries (Persian, Subterranean, Berseem, Ball, Arrowleaf, Crimson, White) and unseeded control were planted October 2017 in tall fescue turf.. Twenty-three pollinator species were observed and 182 individuals specimens. "Background" white clover had the largest observed species count with 72 total. Annual clovers, particularly crimson clover, flowered earliest. In 2018-19, 4 reps (12' X 12') of 5 clover entries and unseeded control were planted October 2018 in bermudagrass turf maintained at 2 mowing heights (2.5 or 4in.). Lower mowing height reduced flower counts for all entries except white clover. Students reported that the experience encouraged their ongoing pursuit of life science research.

[1]Agriculture & Natural Resources Agent, UGA Extension, Whitfield County, Dalton, GA (roger.gates@uga.edu); Students and Associate Professor of Biology, Department of Life Science, Dalton State College, Dalton, GA (ddesrochers@daltonstate.edu).

FERMENTATION CHARACTERISTICS OF COOL-SEASON ANNUAL MIXTURES WITH OR WITHOUT SILAGE INOCULANT

S. L. Shoup, R.B. Muntifering, M.K. Mullenix, S.L. Dillard[1]

High moisture feeds, such as baleage, are gaining popularity due to their consistent nutritive value and timely harvest. The objective of this study was to compare the nutritive value and fermentation parameters of three different cool-season annual mixtures 1) wheat (*Triticum aestivum* L.) + crimson clover (*Trifolium incarnatum* L.; WC), 2) wheat + T-raptor (*Brassica rapa* L. × *B. napus* L.; WT), and 3) annual ryegrass (*Lolium multiflorum* Lam.) + oats (*Avena sativa* L.) + crimson clover (ROC) planted as a 3 × 2 × 8 factorial design (n = 3). Baleage was treated (I) or not-treated (N) with a silage inoculant and sampled at 8 time periods after ensiling. Samples were analyzed for nutritive value and fermentation characteristics at Dairy One (Ithaca, NY). The three-way mixture (ROC) had greater ($P \leq 0.001$) NDF concentration (63.6%) than WC and WT, but there was no difference ($P = 0.100$) between WC and WT (50.6%). There were no differences ($P \geq 0.527$) among inoculant treatments for NDF, ADF, and ADL. Wheat + clover had greater ($P \leq 0.005$) CP concentration than all other forage treatments. The three-way mixture (ROC) had the greatest ($P \leq 0.001$) pH, while WC was intermediate ($P \leq 0.003$), and WT had the least ($P \leq 0.003$; 5.4, 4.9, and 4.5, respectively). Wheat + clover and WT did not differ ($P = 0.140$) in lactic acid concentration (3.7%); both were greater ($P \leq 0.001$) than ROC (0.9%). The three-way mixture (ROC) had the greatest ($P \leq 0.001$) concentration of butyric acid, WC was intermediate ($P \leq 0.021$), and WT was least ($P \leq 0.02$; 2.4, 0.5, and 0.1%, respectively). Results indicate that use of inoculants in these mixtures is not beneficial, but that the mixtures are suitable for baleage production and provide adequate forage quality to maintain animal performance.

[1]Dept. of Animal Sciences, Auburn University, Auburn, AL; sls0017@auburn.edu

FORAGE MIXTURE EVALUATION FOR ORGANIC DAIRY PRODUCTION

S. R. Smith, K. Payne, K. Lea, D. Butler, G. Bates and G. Pigetti[1]

Organic dairy production requires that the milking herd consume 30% or more of their diet from pasture during the growing season. Few farm scale research projects have been undertaken to provide dairymen with forage mixtures, expected yield and quality data, or management recommendations for an organically managed dairy system. The University of Kentucky and the University of Tennessee partnered in a USDA-NIFA-OREI grant to explore several forage combinations and their integration into farm scale organic dairy operations in Kentucky and Tennessee. Combinations ranged from monocultures of cool season grasses to complex, four species mixtures of warm and cool season grasses and legumes. These forage combinations were evaluated with small plot University experiments, ¼ to ½ acre on-farm strip plots, and 10 acre farm paddocks. Mixtures with warm season annuals generally yielded more biomass across the growing season, but cool season perennial mixtures maintained production longer through the season and maintained higher quality during the summer. Not surprisingly, the presence of legumes increased forage quality and maintained higher botanical composition in cool season annual and perennial mixtures in comparison to the warm season mixtures. This research demonstrates that complex forage mixtures provide a beneficial tool for organic dairymen for both yield and quality, but require good management to reach peak production and quality.

[1]Applied Forage Systems Specialist, Virginia Tech So. Piedmont Agric. Center, Blackstone, VA (Kathrynpayne@vt.edu); Research Analyst, Univ. of KY, Lexington, KY (krista.lea1@uky.edu); Ext. Specialist, Univ. of KY , Lexington, KY (Raysmith1@uky.edu); Associate Professor, Univ. of TN, Knoxville, TN (dbutle14@utk.edu); Ext. Specialist, Univ. of TN, Knoxville, TN (gbates@utk.edu); Associate Professor, Univ. of TN, Knoxville, TN (pighetti@utk.edu).

FOSTERING GRASSLAND BIRDS ON OUR WORKING GRASSLANDS

L.E. Judge and M.J. Renz[1]

In the Midwest, grazing and haying begins as early in the growing season as possible to maximize the output of quality forage. Alternative management strategies, however, can produce forage while contributing to the abundance and diversity of native grassland bird species. Their populations have experienced steep declines in recent decades and many are Species of Greatest Conservation Need in Wisconsin. Grassland birds nest on the ground and are vulnerable to the disturbance created by harvest, both machine and animal. Pastures and hayfields provide productive nesting habitat for some of these species when left unharvested until August 1st (the end of the nesting season) or grazed at a low stocking density. These techniques are rarely used, however, because of the expected loss in farm economic productivity. An on-farm study using small plots was designed to evaluate the loss in quantity and quality of forage (using a milk yield calculator) on pastures left unharvested until August 1st. Bird habitat indicators (vegetation height-density and litter depth and cover) were measured to characterize the potential bird habitat and plant community composition was monitored over two years. Results showed that grazed and un-grazed areas within a pasture increased the diversity of potential habitat for grassland birds and that about half of potential forage productivity was sacrificed in paddocks left ungrazed through August 1st. These results underscore the tradeoffs between potential benefits to grassland bird species and the economic sacrifice made by the producer. These impacts to the producer must be addressed in order to reduce or halt further declines in grassland bird populations.

[1]Graduate Research Assistant, University of Wisconsin-Madison, Madison, WI (ljudge@wisc.edu); Associate Professor and Extension Specialist, University of Wisconsin-Madison, Madison, WI (mrenz@wisc.edu).

FROM THE MOUNTAINS TO THE SEA: AN EVALUATION OF NORTH CAROLINA FORAGES

K. P. Phipps, D. D. Harmon, M. H. Poore, A. D. Shaeffer, and J. R. Rogers[1]

Forages are the foundation of grazing animal production. Understanding forage nutritive value can help producers make management decisions that can impact the overall productivity and profitability of their operations. The objective of this project is to evaluate nutritive value and forage production throughout the state of North Carolina and to provide a reference tool for producers and extension agents. Forage analyses reports were collected from the North Carolina Department of Agriculture and Consumer Services (NCDA&CS) Feed and Forage Laboratory. A total of 3,686 samples were analyzed between January 2014 and August 2019. Samples were identified by county, region, species of animal, forage species, forage type (fresh, hay, fermented), and common nutritive value parameters analyzed by wet chemistry. The submitted samples represent 89 of North Carolina's 100 counties with 1,614 submitted from the Piedmont region and 1,301 and 721 samples submitted from the Coastal Plains and Mountains, respectively. The most common species of forage submitted was classified as other grasses, followed by silage corn and ryegrass (955, 259 and 155, respectively). The average TDN for all samples was 62.6%, and 11.6% for CP on a dry matter basis. These results indicate that producers maintain diverse forage production programs while ensuring forage quality remains a high priority in their systems.

[1]K. P. Phipps (kpphipps@ncsu.edu), D. D. Harmon (ddharmon@ncsu.edu), M. H. Poore (mhpoore@ncsu.edu), A. D. Shaeffer (adshaeff@ncsu.edu), J. R. Rogers (jrroger3@ncsu.edu), Dept. of Animal Science, North Carolina State University.

AUTHOR INDEX

American Forage and Grassland Council (AFGC)
PO Box 867
Berea, KY 40403
USA

CURRAN ASSOCIATES INC.
proceedings.com

ISBN 978-1-7138-1448-1
9 781713 814481

Cabin Crew Conference 2019

Identifying and Developing
Professionalism in Aviation Training

Orlando, Florida, USA
30 April – 2 May 2019

Halldale Group